T0231589

Profiting from Property in a Recession

Graham Norwood

2009

Routledge
Taylor & Francis Group

LONDON AND NEW YORK

First published 2005 by Etates Gazette

2 Park Square, Milton Park, Abingdon, Oxon OX14 4RN
711 Third Avenue, New York, NY 10017, USA

Routledge is an imprint of the Taylor & Francis Group, an informa business

First issued in hardback 2016

Typeset in Palatino 10/12 by Amy Boyle
Cover design by Harvest Digital Imaging Ltd

ISBN 978-0-72820-575-8 (pbk)
ISBN 978-1-138-16617-2 (hbk)

Contents

Preface

As a property journalist for the past decade I have been fortunate to see the best of the bricks and mortar business.

Good people have worked hard and made plenty of money. Some lazy people have struck lucky and have made even more money. In recent years, some excellent people have been hit hard in the recession. But housing market downturns are a great leveller. For those brave enough, there is an opportunity to step in and profit from property that has suddenly become much cheaper thanks to the recession.

The emphasis of this book, however, is on responsibility. We are no longer in an era where it is acceptable to urge people to borrow irresponsibly, or simply sit on property and assume it will appreciate. We know how both can end up.

Instead, any money you make on property by following the suggestions of this book will be because you have earned it. This will be done by hard work or by skilful reading of the housing market and good timing. So it is not a plan to get rich quick, just a set of ideas to become better off thanks to your own effort and judgment.

My thanks go to EG Books and my wife Helen Crossfield, to the hundreds of property contacts who have provided the raw materials for this book.

And thanks to you readers, too. Your reward, we hope, will be to make money from property in a recession.

Graham Norwood,
Autumn 2009.

The Market Today

Introduction

"House prices fall 24%", "Properties now back to 2004 values", "Developers desperate to sell stock".

Those headlines have been all too familiar in recent years, and with good reason—the British property market has seen its sharpest falls for three decades and a transformation of mortgage availability.

The price falls are likely to be only temporary, and do little to wipe out the gains we have seen over the past 25 years or more. For evidence of this, look at these figures showing that on average, prices in the UK in the year to April 2009 may have fallen 17.5% but despite this, according to the Nationwide House Price Index, they have risen an average of 520%. In regional terms, the figures look like this:

Region	Fall in year to April 2009	Increase since 1983
North of England	−11.5%	532.3%
Yorkshire	−16.8%	524.0%
North West	−18.9%	484.0%
East Midlands	−18.8%	519.0%
West Midlands	−12.2%	540.7%
East Anglia	−18.0%	525.9%
South West	−12.0%	542.7%
South East	−20.1%	508.6%
Greater London	−20.9%	606.0%
Wales	−18.4%	499.2%
Northern Ireland	−22.2%	629.3%
Scotland	−14.3%	429.7%
All UK	−17.5%	520.0%

There is one simple reason why house prices have risen over the long term in the UK—that is, there are too few homes for too many buyers. Despite the price falls of recent years the long-term demand remains well in excess of supply. So prices will rise again in the years to come.

However, the more restricted availability of mortgage finance will limit that demand more than before, meaning that anyone who wants to make money from property will have to work harder, almost always "adding value".

No longer will people simply buy a property, sit on it and expect to sell five, 10 or 20 years later having made a big profit. That *might* happen, but it is not guaranteed.

But if you add value to the property—make it better, bigger, more useful, with higher quality fittings and more likely to be wanted by others to rent or buy, you massively increase the chance of the home appreciating in value. If you combine that with the fact that right now you can buy at a low price thanks to recent price falls, the potential profit begins to look very attractive.

That is the point of this book.

In the coming chapters we will review a range of ways of profiting from property in a recession, including:

- Self-build—buying a plot of land and building a home on it.
- Re-development—buying a home in bad condition and renovating it.
- Buy-to-let—buying a home, usually a flat, and renting it out.
- International—buying a home overseas and renting it for holidays or long-term lets.
- Commercial—buying a retail, office or industrial property and renting it out.
- Other options—from letting out a room in a home, to buying into a property fund.

One key thing to note early on is that while your ultimate profit will be gained when you sell (and perhaps through regular income when renting out), your first action almost always involves making a purchase.

That is crucial and is what makes it vital to get into property now, after a recession that has made things relatively less expensive to buy.

The underlying assumption of this book is that you wish to make money, not in an irresponsible and reckless way but by serious consideration of whether prices of everything from land to mansions have plummeted in the recession, to allow you to get in now "on the ground floor" and, later, sell at a higher price.

In the interim you must add value—build that house to go on that land, or set to work to turn a wreck into a good quality family house—but whatever you wish to do, the key is to buy when prices are roughly at their lowest. This may well add as much to your profit as all the rest of your work put together, so it's best to do it well.

I say "roughly at their lowest" because, almost by definition, the lowest point of a property price is not recognised until it is gone, and the price is rising once again.

Get a feel for the market

A good guide is to study the many house price indices that exist—we have already seen data from one of the major indices, from the Nationwide Building Society.

It is far too easy to set too much store by a single set of figures from any one of the 15 different price indices produced by mortgage lenders, estate agents, newspapers, property institutions, business consultancies and government departments, but it is sensible to understand what part of the housing cycle each index actually measures.

Now to a casual observer these indices appear confusing or even contradictory. In early 2008, for example, as the credit crunch was biting hard, data from the Land Registry (which collates details of every house and flat bought in England and Wales) showed house prices rising just as indices from the Halifax and Nationwide pointed to falls.

Fionnuala Earley, former Chief Economist at Nationwide, is an expert on indices and says two points must be noted. "First, check where an index is in the 'purchasing cycle'. For example, Rightmove looks at asking prices at the start of a cycle. Halifax and ourselves then look at prices agreed after a buyer's survey and when a mortgage offer is made. The Land Registry releases data only when the purchase is completed, so lags behind a bit," she says.

This means that the Land Registry data released in January 2008 was actually based on prices agreed back in October and November 2007, before the market dipped.

There are other factors to bear in mind when deciphering the statistics in a property index.

A few index authors "modify" data to remove eccentricities such as unusually high prices for trophy homes or low prices for former council flats. This tinkering is done to smooth out irregularities, but because no two indices do exactly the same it makes it very difficult to directly compare figures from different sources.

Comparison of supposedly average house prices for the whole country is hard too, as some indices cover the UK while others only cover Britain (so "low-cost" Scotland brings down the average without being balanced by "high-cost" Northern Ireland). Others just cover England and Wales.

Some indices have PR experts to make sure they reach the widest public, even if they are not always particularly significant. Property data consultancy Hometrack, for example, shrewdly releases its December housing figures every Boxing Day, when there is little other news. Journalists leap on the data, whatever it says, to fill empty columns, making the index perhaps appear more significant than it might be.

If you want mood music and not statistical exactitude, there are "non-price" measures giving a rough barometer of what is going on in the market. The Royal Institution of Chartered Surveyors (RICS) (*www.rics.org*) issues a market survey based on how many of its members (who are actually estate agents) believe prices to be rising or falling, not on prices themselves. However, one point to remember here is that as the property market falls, so fewer homes are sold by agents and RICS members, so their findings are based on decreasing sample sizes.

Then there is the website *www.propertysnake.co.uk*, which monitors property sales websites for price changes. It does not claim forensic accuracy, but at one point in 2008 it showed 118,000 homes on sale, all with reduced prices, in some cases as large as 44%. It will not tell you how much prices have fallen or risen, in the past or present, but it will show you where you might find a bargain in today's market.

The five key indices that attract publicity are:

- **Land Registry** found at *www.landreg.gov.uk*, and based on a large sample of completed sales, is regarded as the best. But as house purchases routinely take three months to complete, it is slow to appear so is not helpful if the market is volatile or just on the turn.

    The Registry does not predict what will happen, preferring instead to report on what actually has happened. So, by early summer 2009 it was saying prices in England and Wales had fallen 16.5% in the past year but that the average price of a home (£153,862) was still 246% of what it was when the Registry began publishing this type of information in January 1995.

- **Halifax Bank of Scotland** found at *www.lloydsbankinggroup.com*, and based on a sample of mortgage approvals, is much quicker to appear. But it is stronger in some regions than others—in central London, for example, most house buyers do not need mortgages.

  By early summer 2009 it was saying that the volume of sales in England and Wales halved between 2007 and 2008, but there were some very tentative signs that activity may be beginning to stabilise. The house price to earnings ratio (a key measure of housing affordability) was at its lowest level since early 2003 at 4.34, suggesting homes were more affordable again for the likes of first-time buyers. But it warned that increasing unemployment, low consumer confidence and the continuing volatility of the financial markets on the availability of mortgage finance would prevent dramatic price rises for some months to come.

- **Nationwide Building Society** at *www.nationwide.co.uk*, again based on a sample of mortgage approvals, is the oldest monthly index—now in its 35th year—so it has a vast reservoir of trend information to draw on for every part of the UK.

  By early summer 2009 it was saying that prices in southern regions of the country were more sluggish than those in the north, with large falls still being recorded in Scotland and Wales, but Northern Ireland—which had suffered big price drops—were beginning to moderate. The amount of interest from potential buyers was at its best since spring 2008 but that, against a backdrop of national economic gloom, it was too early to talk of a house price recovery.

- **Hometrack** at *www.hometrack.co.uk* relies on data from 4,000 estate agents and was the first to predict and quantify the recent slowdown in both prices and transactions. It also works closely with developers, agents, councils and housing associations.

  By early summer 2009 it was showing house price falls at their most modest for 10 months because of what it called "pent up demand" after such a long period of slow sales. But it warned that with increases in unemployment and weak economic performance across the economy being anticipated, it doubted that prices or sales volumes would rise significantly in the near future.

- **Rightmove** at *www.rightmove.co.uk* is based on asking prices of homes advertised for sale on the *Rightmove* website. It accounts for about a third of all homes on sale so is representative, but because it reflects only asking prices, pushy estate agents and greedy vendors can make the market appear more bullish than it is. However, it gives a thorough snapshot of what sellers hope to get—even if they fail.

By early summer 2009 it was reporting "renewed buyer interest" now that prices were about 25% below their 2007 peak, with many vendors who did not absolutely have to move being deterred by the prospect of selling at such a discount. Occasionally, vendors even increased their asking prices—with the help of estate agents—hoping to kick-start the market.

Other useful price indices can be found on: *www.communities.gov.uk*; *www.ros.gov.uk*; *www.knightfrank.co.uk*; *www.savills.co.uk*; *www.cbre.co.uk*; *www.cluttons.com*; *www.primelocation.com*; *www.acadametrics.co.uk*; *www.rics.org*; and *www.naea.co.uk*.

Although by the start of summer 2009 there was no firm evidence that prices were picking up, careful analysis of the five major indices shows that there were the first signs of greater interest from buyers—an indication that the bottom of the market may not be that far away.

Do local research

This broad brush national research must be complemented by research at local level too:

- Talk to local estate agents about key streets, favoured house types, local issues and the nature of sellers in the area, even if you do not intend to buy a property that they have on sale.
- Scour websites such as *www.mouseprice.com* and *www.house prices.co.uk* which show recent local sales and prices—these can provide hard evidence to show an estate agent and seller when you make a bid on a property.
- Check on schools, transport points, road noise or other issues which may be pertinent to the kind of property you intend to buy, build or modernise.

One word of warning, however—don't believe everything you read about property markets in local newspapers. Although the property supplements in the national newspapers are written by independent and skilled journalists, often the complete reverse is true in local newspapers where the free homes-for-sale supplements are often written by estate agents directly, who are unlikely to give an independent assessment of the local trading conditions.

Be a hard-nosed buyer

For many, this is the most difficult part of making money from a property—harder even than harnessing large numbers of workmen to renovate a house, or finding tenants to pay your rent. You make the most money from your property if you start by paying the least you can for it. So be tough, and back up your position by careful research and evidence.

Researching the reason behind the seller's decision to move is vital. Ask the estate agent the following questions:

- Why is the vendor selling? (The answer to this may reveal whether the seller needs to get out quickly and will accept a low offer. Sellers who are divorcing or need to move to get a child into a new school catchment area are especially flexible on prices.)
- What is their timeframe? (ditto.)
- How long has the home been on the market? (The longer it is, the more desperate the seller may be, depending on circumstances— or it may tip you off that the seller is completely inflexible on price, in which case walk away.)
- Have there been previous offers? If so, what were they, how did the sellers respond and why did the deal not go through?
- Has there been a survey conducted of the property, and if so what did it reveal? (Estate agents are notoriously prissy at answering this one, but be persistent as if there is a major problem, it may be revealed this way at no cost to yourself in terms of time or paying for another survey.)

Adopt a rigorous checklist to look at everything you can within a property.

The checklist below has been compiled from information from a series of well respected buying agents. Their job is to act for well-heeled clients who want to purchase a property to live in or for investment; the agents then shortlist suitable examples and, when the client has chosen the one he or she wants, the agent negotiates the best possible deal. Buying agents can be ruthless so look for every possible justification to reduce the offer to the seller.

Inside
- **Kitchens**: Are there leaks near the sink, dishwasher or washing machine? Check the ceiling for marks as the kitchen is usually just below the bathroom. Also check the water pressure is OK in the taps, particularly if the kitchen is on an upper floor or in a high-rise apartment.

- **Bathrooms**: Get in the bath and check if it then separates from the seal. If the boiler is new, see a guarantee. If not, ask about service history (it must be checked annually). If it's old, cut your offer by the cost of a new one.

- **Bedrooms**: If you see blankets, hot water bottles and so on, the room is cold. Why?

- **Living rooms**: If there's a wooden floor, check to see whether it's noisy. And does it contravene the lease in a flat? Ensure carpets and lights are included in the sale if you want them. Where there's a chimney check the brickwork, black marks suggest the flue is damaged.

- **Staircase**: Send someone upstairs and check for vibration or weakness.

Outside
- **Brickwork**: Any cracks? Is repointing required?

- **Windows**: Ask for guarantees if they have been replaced.

- **Roofs**: Are they sound? Slate tiles usually last 15 years and a flat roof needs replacing after just seven years.

- **Gutters**: If they're blocked, water may well have penetrated the brickwork.

If you see a problem but still want the property, set an ultimatum—perhaps get the vendor to pay for and fix it. If the seller refuses to play ball on this, it may mean that they are strapped for cash so you can negotiate harder with a reduction to compensate yourself for having to do the work should you buy it.

Case study: Lana Wilson

Negotiate hard is exactly what public relations consultant Lana Wilson did when she bought a four-bedroom Victorian end-of-terrace house in Hackney, London, in February.

"I short-listed three streets in Hackney where I wanted to live, and as a house there came on the market I checked its historic prices with Land Registry data. Then I spoke with different agents to see if they thought the price was fair," she says.

She viewed each property, finally putting in an offer on one home that was well below its £900,000 asking price.

"My offer was rejected so I just walked away and didn't try to negotiate. Two months later the estate agent came back and said the vendors were interested. But by that time prices had fallen further, so I recognised the sellers were probably desperate to move. I finally got it for £750,000, which was only a little more than it sold for several years ago," she says.

Wilson also did two things that are considered critical in securing a lower than expected purchase price—she researched nearby sales and sorted out her finances so she could faithfully tell the seller she was ready to move quickly.

Why will prices rise anyway?

This is a good question. For over two years between summer 2007 and late summer 2009, a price rise has not looked likely, so why should it happen in the future?

Well, the best evidence within the UK comes from an influential economics consultancy which says that house prices have to go up for one simple reason—government targets for the minimum number of new homes are just not being met.

Look at it this way. In 2006 "targets" were the order of the day and Gordon Brown announced that in England alone there must be between 240,000 and 297,700 homes built annually until 2016. That means that, in total, between 2.9 million and 3.5 million new homes must be built by 2020. In 2007—before the downturn hit the new-build market—some 174,900 homes were completed: still below target but on an upward path from previous years.

Yet in 2008 the total number of housing starts is believed to have been scarcely 100,000 and for 2009 and 2010 the totals may be merely a dismal 55,000 to 75,000 annually. The consequence is that the new homes industry is imploding. The fewer homes built, the fewer people work in the industry, making the downturn even worse. Similar targets, and shortfalls, exist in Scotland, Wales and Northern Ireland.

Jim Ward of estate agency Savills predicts new-build levels in England will not have returned to 140,000 per year—scarcely half the Government's target—even by 2013. He says:

> Once sites are mothballed, there's inertia in the system as it takes time to rebuild teams and return to the same master-planning position on more complex sites.

Recent measures announced to kick-start elements of the housing market have done little if anything to improve building rates. Stamp duty changes may boost sales but not new building. Allowing housing associations to buy unsold houses and flats mops up stock but does not help overall supply meet demand.

The slump in building will not just affect the private sector. Many of the affordable homes made available to key workers and the low paid are built as a result of so-called "planning gain" deals—that is, arrangements by which developers have to build a number of affordable homes to get an agreement to build private sector properties.

With private sales at a standstill, developers have downed tools on residential property across the UK, so "planning gain" properties are not being built either, creating a shortage of properties exactly where demand for affordable homes is strongest.

That fundamental mismatch of supply and demand, of course, will fuel a new housing boom if the Centre for Economics and Business Research's (CEBR) forecasts are correct.

"The sharp drop in completions will mean higher prices if and when credit markets sort themselves out" says CEBR Senior Economist Ben Read:

> The government will be concerned that, with every year that passes, it gets further away from its targets. With developers unlikely to respond quickly when the market bottoms out, prices may recover more quickly than people imagine.

That price rise may be every bit as large as the price falls seen during the recession.

A report published in 2008 by the National Housing Federation (NHF) suggests the average house price in England will rise 25% by 2013. NHF Chief Executive David Orr says:

> House prices will increase substantially over the mid to long term. Demand is going up, while the supply of new homes is going down.

That, simply, is why I think this is a good time to profit from property in the UK.

Overseas is a different story. Few countries have the apparently permanent under supply of homes that exist in the UK and if you buy anything that you wish to use yourself—from a sunny coastal villa to a *pied-a-terre* in a European capital—you will need to make a choice based on your personal desires as much as on what will maximise your profit.

Remember the long-term

Here is one final thought before we move on to look at specific ways of making money from property in a recession.

If you want to buy bricks and mortar and get a profit in just a year or two, put this book down now. One lesson of the recession and of property investment, learned by serious investors many years ago, is that the biggest gains are over the long-term.

Research by Savills (using Nationwide data, as we did above) shows that on average, every 10 years between 1968 and 2008, UK residential property has grown in value by 170%. That means that during the past 40 years prices have grown on average by 2.9% per annum, adjusted for inflation.

So think long-term.

Resource centre

In addition to the websites listed above, have a look at these.

The most popular sales website for homes priced at £500,000 or more is *www.primelocation.com*, while *www.periodproperty.co.uk* is, as you can imagine, the place to find a thatched, listed or historic home. This site—*www.firstrungnow.com*—targets first-time buyers and has information on low cost properties for sale.

The website *www.nestoria.com* has top notch data from car parks to pubs, while *www.ukvillages.co.uk* is good for maps and aerial views of where your new home is. Whatever you buy or sell, you will encounter a Home Information Pack. See what you're in for at *www.homeinformationpack.gov.uk*.

Which?, formerly the Consumers Association, has a guide for inexperienced buyers and sellers on *www.which.co.uk*, including a model contract and tips on dealing with agents. If you run into problems, complain via the Office of Fair Trading (*www.oft.gov.uk*) or the Ombudsman for Estate Agents (*www.oea.co.uk*).

Self-Build

Introduction

If you think this option sounds like hard work, you are right—but the rewards are enormous.

Self-build sounds the stuff of anoraks who like building sites and spirit levels. But it's not as rare as you might think and is done by all sorts of people, many with no experience whatsoever of construction.

Before the recession, around 20,000 people built their own homes in the UK each year, providing about 12.5% of all new homes that were built annually. That proportion is small compared to other countries (in some parts of Scandinavia and central Europe about 40% of all homes built are by individuals).

The UK proportion is likely to increase markedly in the years to come as various reports and surveys commissioned by successive governments have urged tax breaks and financial incentives for these kinds of development.

Those will come in the future but even at the moment, it makes fantastic financial sense to try this approach, and historically the self-build entrepreneur has fared well in downturns. The National Self-Build Association says:

> In previous downturns the self build market has fared well. Indeed, in the early 1990's, when house prices last slumped, there was a significant upturn in self build activity. This was driven by the fact that home owners realised that they could make the value of their existing properties work better for them if they sold up and used the money to

fund a self build project. A recent survey of 500 would-be self builders reinforced this message, with nine out of ten of them saying they would still be undertaking their projects, despite the country's current economic problems.

However, before you start on your own project, you need to think very seriously.

As a property journalist, I have reported on literally scores of self-build projects. Contrary to the dramas you can see on Channel 4's *Grand Designs*, most do not run over budget, nor involve monumental disputes with family members and/or builders, and there are very few catastrophes.

But they do all have two things in common—they involve immensely hard work and personal sacrifice.

The hard work is obvious; even if you hire in expert builders you will still need to supervise, order materials, project manage and do the books, plus negotiate deals on everything from the plot of land to the mortgage and the bricks and mortar.

Yet, the hardest thing is quite different. It is that, because most self-builders have to sell an existing home to fund the project, they end up living in a caravan on site, or sharing accommodation with friends—this arrangement can easily last one year or longer. What seems like an adventure when first discussed can wear very thin when you wake up for the 300th day in a row in a chilly caravan on site, next to a partner and perhaps even children who are similarly disenchanted.

The darkest hour comes just before dawn, and despite these privations most self-builders carry on to complete their projects knowing that, however depressing their surroundings, they will sooner rather than later end up with their ideal home and a good profit to boot. But take these things seriously, discuss them with your family and have a good long think before embarking on such a project.

Pounds and pence—why self-build?

Building and then selling the completed product is usually designed to be a profitable exercise and homes are no exception.

There are several immediate reasons why self-build can be very profitable:

1. Value added tax (VAT) is zero-rated on self-build homes so any VAT incurred on building materials, labour employed on site or

fixtures and fittings can be reclaimed. This is a one-off process as we explain later, and it can only be done when the property is completed.

VAT rates vary in the UK, and during the height of the recession the most common rate was temporarily dropped to 15%. So, say, you spend £115,000 on materials, labour and decorations; under this temporary VAT period, you would be able to claim back the £15,000 at the end of the construction.

Even as VAT levels change, as they will when the recession ends, the same principle applies.

2. The only stamp duty paid by the self-builder is on the plot of land, not on the overall value of the property that is constructed. This is unlike, say, redeveloping an existing property for profit, where you may have to pay stamp duty on the original purchase.

    Again, stamp duty was temporarily changed during the recession, but the "usual" levels were that stamp duty was levied at 1% on property sales from £125,000 to £250,000, 3% for £250,000 to £500,000, and 4% for properties selling for above £500,000.

    This means if you buy a plot for £155,000, and the final value of the property is £525,000, you would pay stamp duty of £1,550 (1% of the plot cost). In contrast if you bought a completed home for, say, £275,000 before you improve it, then you would have to pay £8,250 stamp duty (3% of the property cost).

3. You can make a larger profit margin than a professional developer. Commercial developers, say the Home Builders' Federation (the umbrella group representing the majority of large construction firms), achieve a 15% profit margin on each property. So if a home is sold for £200,000 about £30,000 of that is profit for the firm.

    Although the margin on a self-build can vary, by doing as much work as possible yourself, you could make at least double that figure, and may, depending on the individual circumstances, make even more.

4. The recession is an ideal time to self-build because land prices are low (they fell around 35% in 2008 alone and another 5% to 10% in the first half of 2009) and because there is plenty of labour available to help your construction project.

Step 1: Budget

At first you will not know for sure exactly how much anything costs—or, for that matter, even the exact location or price of the plot you wish to buy. But you need at an early stage to consider "outline" costs in order to arrange a mortgage of the right type and size. The more you have worked out, the more precise your approach to the lender will be—and the more likely you are to get the loan agreed.

There are several ways to work out how much you need to spend on your build.

You could use an Excel spreadsheet and make *frequent calls* to a builders' merchant to work out the pricing. Alternatively, you could *buy professional builders' price guides* costing up to £75 that do the job themselves.

These approaches are very time consuming, so to help make the pricing process easier, a number of companies have introduced *computer software packages*. However, many cater specifically for building professionals, so you'll want to choose from those aimed at self-builders. Before you rush out and buy one, it's worth checking the products on the market.

Get a good one and your package will estimate labour, materials, time and costs for full or part houses, extensions, garages or outbuildings of any size. Generally, the software will put the estimates on all common PC word processing and spreadsheet packages, or can directly print off-screen.

All you have to do is feed in dimensions, phasing, chosen materials and variables, such as customised labour costs, and state whether you are doing any of the labour yourself or using subcontractors.

You enter the costs of the items you need. If you are unsure of the quantity, the packages help calculate this by entering the cost of an item and the space or volume to be covered. If, for example, you enter the cost of labour for glazing one window then the number of windows you need, it does the calculation.

The software packages make calculations based on broad assumptions—for example, that the ground is level so only straightforward foundation work is required—and no allowances are made for weather or delivery delays unless you build in a contingency.

If you are nervous about installing software on your PC or don't want to risk a high outlay on a computer program you may not be able to use, there are third-party estimation services. In these, you submit drawings, either on paper or from an architect, and include details of likely labour rates (low if you are employing friends or family), or local rates from professionals you must approach before submitting your details.

When calculating your budget, remember to take account of:

- "hidden" costs like planning application fees, site insurance for the duration of the build, fees for checking building regulations and so on
- your own living costs (whether it is in rented accommodation or your existing mortgage on your home) as well as project costs
- advance payments necessary for some materials and contractors.

Expect to pay the following for basic fees as you pursue your project:

- Up to £1,000 in arrangement fees if you change mortgage lender from your current "bought" home to a different lender's self-build product.
- Up to £300 for a survey of your plot.
- £450 for agency and Land Registry fees when you buy a plot—on top of the cost of the plot itself, of course.
- Planners' fees for resubmitting drawings can be £1,000.
- £2,000 for insurance.
- £1,500 for connection fees for utilities.
- Inspections for building regulation compliance may cost £350.
- Furniture storage if you sell your old home could be £200 monthly.

At this stage you are probably wondering why anyone ever builds a home, but remember that 20,000 a year do so—and if they can manage it, so can you.

It sounds complicated but it isn't. It is absolutely worthwhile, as it not only gives you good financial estimates but it makes you focus on the type, size and quality of the home you want to build.

Step 2: Getting a mortgage

This used to be relatively easy, as we all know, but now it is less so as a result of the credit crunch.

Of 2,915 mortgage products available in summer 2009, 12% or 355 are for self-builders. Deposits vary widely but were always high. For a property with a completed value of £500,000, most applicants had to find a minimum deposit of 35% or £175,000.

Now it is still possible to get specialist mortgages for up to 95% of land purchase and for the cost of a timber frame kit property or indeed the "true" architect designed self-build. However, it is increasingly common that a maximum of only 75% is available, meaning you almost certainly will have to sell your old home before starting the project.

The unusual characteristic of self-build mortgages is that most will pay only in *stages*—usually in arrears, meaning you may have to also arrange a bridging loan to cover the cost of each stage before repaying it once the phased mortgage funding comes through. There are usually six stages to self-build mortgages:

1. Paying for the land—the costliest phase.
2. Building the foundations.
3. Contingency borrowing (often about 10% of the total loan) for emergencies.
4. The main construction phase.
5. The final waterproof sealing of the property.
6. Completion of the project.

At each phase the lender will require proof of you having reached this stage of the build, and that what you have done is of a suitable quality. Often this will entail a local planning authority signing off your work—sometimes the lender sends its own representative.

Remember that although the payment process from the lender to the borrower is very different than when purchasing a "built" property, the variety of *loan types* is roughly similar.

The loan types are: *fixed rate mortgages*, which do not change their costs over a specific period, irrespective of what happens to interest rates. A pro is that you will not be caught out by unexpected interest rate rises, but a con is that you do not benefit from falls in interest rates.

Discounted variable rate mortgages give you a discount on the lender's standard variable rate for a specific period. The pro is that you have low payments during its period, so it can be helpful if you are spending money on, say, an expensive part of the build or in furnishing the finished product. The con is that when the discount period finishes, the payments you make each month may cruelly bounce to a high level if interest rates have risen in the interim.

Tracker mortgages track the Bank of England base rate and when the base rate changes, so does the amount that you pay, while capped mortgages set minimum and maximum amounts you pay, whatever happens to be the "mainstream" interest rate in the wider marketplace.

Self employed people in particular used to be able to get self-build *self certification* mortgages but these are extremely rare now and are unlikely to be widely available in the future.

Step 3: Finding a plot

There are several ways of doing this—but first, a warning.

Remember that there are many firms now selling land online that has not got, nor is likely to get, planning permission for housing. They sell plots at up to 10 times their agricultural value but claim it may one day be allocated for homes. Most analysts consider these to be unwise investments.

But hundreds of legitimate plots do exist across England, Scotland, Wales and Northern Ireland. However, time is of the essence as in a recession—especially our current vicious recession— professional property developers are not buying land so it is much easier for "amateur" self-builders to find an ideal location.

So how do you find a plot?

One way of finding them is by looking on one of the respected *internet plot-finding directories* which specialise in "small ads" from individual sellers and from big developers who have end-of-scheme sites which they want to dispose of.

At the depth of the recession I looked on one of these, narrowing my search only to the East Anglian region. I found no fewer than 264 plots in Cambridgeshire, 254 in Lincolnshire, 207 in Norfolk and 85 in Suffolk. About 75% of these had planning permission for residential development but 25% did not.

Those with planning permission are obviously more expensive than those without. Land owners routinely get permission for their plot, even though they intend to sell the land and never intend to build on it themselves—but by having residential planning permission, they hike up the site's value and give you a firm indication that you could build on it. If you consider a plot that has not got such permission, contact the local authority planning department and/or a local private planning consultant to ask whether a residential property is likely to be permitted on it.

A further way of locating a plot is to *put an advertisement in a local paper or community website* saying you want to buy a plot, and browse local media classified sections for advanced notice of planning applications—these will tip you off about plots likely to come on the market shortly.

More pro-active self builders will choose the area in which they want to build and then make an intensive study of existing homes. If some have grounds that are large enough to accommodate a new home and provide both access and privacy for all concerned, it is just possible that those owners will be happy to sell a part of their garden. So you could *leaflet homes with large gardens* that fit the bill.

A variation on this is what self-build experts call "bungalow bashing". This is buying a bungalow that has seen better days and demolishing it, releasing the large plot of land that it sits on.

A key advantage of this tactic is that you stand a strong chance of getting planning permission for another property to take its place. It is likely that drainage is good and electricity, water and gas are already available. It is also probable that the land will not be contaminated through previous industrial use or by being treated as a dump—a common problem with brownfield land that has previously been used for non-residential building.

The disadvantage is cost. Obviously you pay for the land and the property, even if you knock down the latter. Some canny owners know that self-builders are keen and add a premium to their asking price, so if you try this approach you should be in the mood for some tough negotiations.

A further option is that if you are willing to live near an existing housing estate or development, you could *contact commercial developers* who are active in your chosen area. Developers have large landbanks, often stored for over 10 years, and will sometimes dispose of small tracts which are no longer required.

Other options include contacting the *Society for the Protection of Ancient Buildings* (SPAB) for details of properties requiring renovation or conversion. This is not quite the same as self-build, and SPABs' lists usually consist of listed buildings which can, frankly, be difficult and expensive to renovate.

You can also try your hand at a land auction, but be warned—property auctions of any kind can be quite intimidating.

If you wish to go to a land auction, prepare in advance. Most auctions are in hotels and are advertised by participating estate agents and valuers. Contact them for a catalogue of plots under the hammer at the next auction (often this is done via a premium rate telephone line, costing about £5 a call).

The catalogue will contain photos of the plots, details of condition and any planning permission, and guide prices, which are normally the minimum the sellers will accept.

If you like the look of a plot, visit it well before sale day. Auctioneers arrange group viewings for all potential buyers or ask local estate agents to do individual viewings.

Some sellers and their estate agents have "pre-sale surveys" of properties and you can do this on a plot too, to ensure it does not have contamination and to get some indication about whether local planners will agree to a home being built on it. You may have to pay between £200 and £500 for this.

Ahead of the auction, arrange a solicitor and mortgage lender as in a conventional house purchase. Ensure you have funds covering the likely sale price (experts suggest the guide price plus 15% to be safe). Usually you must pay 10% of the property's cost at the auction if your bid is successful.

At the auction each plot is called a "lot". When your preferred lot comes up, the auctioneer will confirm the address and details of the plot before asking for bids. The bids usually rise in £5,000 levels until they approach the guide price, when bids will rise in £2,000 or £1,000 sums.

Stay cool and remain well within your budget, or even get a friend to bid on your behalf if you are nervous. If your bid is successful, you pay 10% of the cost immediately and the rest within 28 days. Remember —once the hammer has come down, neither the buyer nor seller can withdraw.

In addition to any or all of the above, you can also simply *register with local estate agents* in the area you want to build.

Although obtaining a plot—through whatever route—is only the first step in many to constructing your home and making an eventual profit, it is likely to be one of the most costly steps you will take.

Prices vary enormously but a plot typically costs 40% to 45% of the overall price of self-building a home. In the South East of England, where all prices are more expensive, that soars to between 50% and 60% because of the scarcity of land, or possibly as much as 70% if you buy an old home and knock it down.

Case study: Paul Gladman and Kerry Medhurst

Paul Gladman and his partner Kerry Medhurst were too impatient to wait for a plot of land to come on the market—and that's just as well, because otherwise they may not have built their award-winning five-bedroom, three-storey, self-build home in the Essex village of Danbury.

"Rather than wait for a piece of land to appear we spotted a 1920s bungalow within the village envelope. It was a small house with a large plot but we had to move fast. We knew professional developers could be after it", says Paul. "The plot was 65 ft wide by 300 ft long, and, with the bungalow, cost a cool £250,000."

"My advice for other self-builders is don't get disheartened. It can take a while to find a plot or a bungalow to demolish, but they must be absolutely ready to respond when it appears. Have the funding arranged and, if possible, talk to planners to see if your proposals are likely to get planning permission," says Paul.

Step 4: Kit or architect?

Your next choice is to decide whether to build a unique, architect-designed property or a kit house.

Kits cost up to £125,000 before construction costs, but designs may lack the personality many self-builders want. Research by the housing think-tank, the Joseph Rowntree Foundation, shows that self-builders want features that cannot be found in standard homes. They look for a room in the loft, basements, energy-saving features such as water recycling units, or technical gizmos such as under-floor heating.

It is therefore perhaps likely that you will instead *commission an architect* to design a home incorporating all your needs—but fees for that may be as much as £20,000.

Many novice self-builders are intimidated by the language and costs of architects, but if you shop around you do not need to be frightened.

A self-builder should write down what they want to achieve—say, number of bedrooms, look and feel of the building, kinds of materials and so on—and give the architect a firm indication of the overall budget for a property. The architect should then think laterally and come up with ideas which novices alone might not have considered.

A good start is to find an architect who has experience of self-build projects and to speak with the architect's clients to see how things worked out. Try this guide to finding and working with an architect:

- Identify an architect through the Royal Institute of British Architects (RIBA) with self-build expertise. Remember to be in the RIBA to begin with they will have a degree in architecture and have passed professional practice exams. (Tel: 020 7580 5533, *www.architecture.com*).

- Set up an initial meeting to discuss your project. This may be free, even if you do not commission the architect, or it may involve a small fee.

- If you wish the architect to do more work, request an itemised cost breakdown for the architect's work—remember, there is no standard fee rate and charges vary from practice to practice, so shop around if you wish.

- Once commissioned, they should then offer you preliminary project advice and prepare drawings, taking them to the local authority planners for consent.

- If you wish, and for additional fees, the architect can obtain tenders of work from contractors, help you choose the best, prepare relevant contracts and write a schedule of work and a timetable. The architect can even manage the project on your behalf if you can afford it.

Step 5: Manage it yourself or hire a project manager?

The ultimate self-builder will give up their job and manage the entire project from start to finish. It gives a tremendous sense of satisfaction and may cost "only" your salary of, say, £30,000. This is compared to the likely cost of up to £50,000 for a year's work with a project manager.

It sounds a no-brainer to do it yourself, especially if you want to maximise your eventual profit. But ask yourself one question—can you really do it?

There is no shame in admitting you cannot, and there may be much wastage or time, money and patience if you try but fail, and then call in a professional anyway.

If you want to do it yourself, be sure you're happy to talk to all of these people:

1. **Architects**: as we have said, you will need a chartered architect if you want a truly unique property, or at least the help of an architectural technician who can produce drawings to accompany your planning application and to satisfy council officials who are monitoring building regulations.

2. **Bricklayers**: these will require on site instruction and supervision.

3. **Building control inspectors:** for most projects these make nine separate visits to ensure the building work conforms to legal regulations, and they must also sign off the completed project. Their support is usually essential to trigger payments from your self-build mortgage lender and they are particularly "hot" on building regulations which are split into "parts":
    (a) Part A: structural safety
    (b) Part B: fire safety
    (c) Part C: site preparation
    (d) Part D: toxic substances
    (e) Part E: sound insulation
    (f) Part F: ventilation
    (g) Part G: hygiene
    (h) Part H: drainage and waste disposal
    (i) Part J: combustion appliances
    (j) Part K: protection from falling and impact
    (k) Part L: energy efficiency
    (l) Part M: access to and use of land
    (m) Part N: glazing
    (n) Part P: electrical safety.

4. **Councillors**: you may need to lobby the elected district or county councillors in the area where you build if your planning application is in any doubt, or if there is any likelihood of objections from existing residents.

5. **Electricians**: ensure they have the appropriate qualifications.

6. **Engineers**: a chartered engineer may be required to establish the type of foundation necessary.

7. **Estate agents**: you will need to register with these when seeking your plot, and if possible encourage them to give you advance information on sites coming up.

8. **Groundworkers**: these build the foundations and may be involved in installing drainage too.

9. **Insurers**: you may need to insure your site against contamination or vandalism during the build; your craftsmen against injury, and your completed property with one of the traditional new home warranty operators, such as the National House Building Council.

10. **Joiners**: these will be required for floor and roof timbers, doors, staircases and outbuildings.

11. **Lenders**: establish an overall agreement and the total mortgage then liaise at intervals, providing evidence of satisfactory completion to trigger threshold payments.

12. **Neighbours**: courtesy and ensuring you do not get off on the wrong foot requires you to introduce yourself to future neighbours and keep them up to date with your plans.

13. **Planners**: on a virgin site you may need separate outline and detailed planning permission, and to consult before the first design is drawn and several times during building.

14. **Plasterers**: ensure they can render, bond and skim, and in some cases dry line basements or rooms near damp ground.

15. **Plumbers**: these are expensive, so provide clear instructions and appropriate monitoring.

16. **Roofers**: general craftsmen can lay modern interlocking tiles; use skilled roofers for clay or thatch.

17. **Scaffolders**: these are also expensive and work under strict health and safety regulations, so you will need to select a skilled company.

18. **Suppliers**: you will need to speak to at least one builders' merchant (possibly many more) and use your charm and customer skills to seek bulk order reductions and prompt delivery.

19. **Utility firms**: gas, electricity, telephone and cable companies will all require contact and co-ordination.

20. Many self-build experts say that if you are in doubt about your ability to communicate and manage resources over a lengthy period, then you should hire a professional project manager. The costs can be high—up to £30,000 for six months in exceptional circumstances—but this can avoid wastage if you doubt your own ability to communicate.

If you do want to hire a project manager, what do you do?

If you are building a kit house, the answer is easy—hand it all over to the supplier, whose staff will be expert in this relatively simple project management task. Or you can pay your architect to manage the process—the architect, too, will be skilled but you may want to question how much of your profit you will lose by hiring such an expensive professional (who is dearer, per hour, than most project managers).

Instead, you can find a project manager by scouring websites on the *internet* or by contacting *professional bodies* like the Project Management Institute (*www.pmi.org.uk*) and finding individuals experienced in handling self-build projects like yours.

Step 6: Build it

Getting the right builders is a crucial part of the project for yourself or your project manager.

Check with reputable online *self-build directories* or the *Federation of Master Builders* (*www.fmb.org.uk*) for quality contacts. Get quotes with guaranteed start and end dates. A big cause of overspending is time slippage, during which you pay the builders and your own accommodation costs or a bridging loan if you are living in a conventional house.

Which?, the advice division of the Consumers' Association, has helpfully set out a list of questions that you should ask prospective builders when you are looking for the right people for your project:

- Do they belong to professional organisations?
- What references can they give you?
- What qualifications do they have?
- What insurance do they have?

- Can they provide a detailed quote setting out the cost of materials they wish to get on your behalf (if any), the likely time to undertake the work and a fixed cost for the project or at least a daily or hourly rate?
- Are they willing to give timing, costing and quality guarantees in writing?
- Are they happy to enter into a contract with you?
- How do they want paying for the work, and in what phases?
- Do they understand current building regulations?

Some things are painfully obvious. Do not hire the first builders you see and do not pay the whole cost upfront.

What is less obvious, for example, is whether they or you should handle subcontractors—bricklayers, groundwork and clearance workers, carpenters and joiners, plasterers, electricians, gas installers, plumbers, landscapers and so on. If it is down to you to select them, apply the same principles as set out above—and remember, when you consider how much work is involved, whether it is actually more sense to employ that professional project manager from the outset.

Step 7: Completion

Most self-builders speak ecstatically (and exhaustedly) of the euphoria they feel when the last brick is in place. But there is one further task before putting the home on the market—it's bureaucratic, but worth the effort.

You can reclaim the VAT so long as the property is clearly a private domestic residence and that you have the paperwork covering your transactions and VAT payments. You cannot claim a refund for the VAT on some professional fees, on the hire of plant or scaffolding, on delivery fees, but otherwise much of your VAT will be returned. The claim must be made within three months of the self-build being completed and you cannot "go back for more"—so get it right first time. Follow advice from Her Majesty's Revenue & Customs (HMRC) on *www.hmrc.gov.uk*.

Then prepare to sell. After all your hard work, remember to keep yourself professionally "objective" about the home, and don't be seduced by a smart-talking estate agent who may sound good but lacks a track record of success.

Before selecting an estate agent, you should get valuations and advice from at least four agents. Try these tips:

- Scrutinise similar homes on sale nearby to ensure an agent does not over-value your property.
- Ensure agents do (or do not) accompany viewings, whichever you prefer.
- Ask agents how long they took to sell similar homes.
- Request details of recent clients and ask them how agents performed.
- Check that there are no extra fees for advertising or website promotion.
- Reject "tie-in" contracts that forbid you dropping an agent or switching to a rival.
- For sole agents (where only one has exclusive selling rights), fees are usually 1.5% to 2% of the eventual sale price.
- For joint agency (where two have shared selling rights), you pay 2% to 3%.
- In rare cases of multiple (three or more) agents the fee is 2.5% to 4%.
- Remember that some agents routinely expect sellers to pay one-off fees for advertisements, glossy brochures or promotions.

Case study: Kester and Zinnia Wilkinson

Self-build has always been popular with those pioneers who wanted energy efficient homes. This was the case even decades ago—now, with everyone talking about the environment, even more are looking to self-build to achieve their dreams. But beware, not everyone shares green enthusiasts' passion for eco-features so check that your eventual cost will still allow you to turn a profit.

Kester and Zinnia Wilkinson from Hertfordshire stumbled across the idea of self-building back in 2000, when they read a book on the subject. From then, the couple, and their daughters India and Liberty were on the road to building their own home —but with an eco-twist.

"I was researching building materials when I was recommended to go on a straw bale course," says Kester. "Then people kept coming up to me saying there were more environmentally friendly ways of doing other things. They introduced me to materials such as lime and breathable paint. Before that, I'd been into things like chipboard."

Their two-storey property, finished in mid-2002, is timber framed and has foundations filled with vermiculite, a sponge-like mineral that provides extra insulation. The walls are made of straw covered in clay, with a plaster finish, and the roof is shingle covered.

The Wilkinsons spent £100,000 buying the land and knocking down a 50-year-old bungalow that was on it, and then another £70,000 building the house—£20,000 more than planned.

"But it's a false economy to hold back on eco-products," says Kester. "Using breathable paint at £6 a litre, for example, sounds expensive. But it requires only one coat and stays in good condition longer than conventional paint. The long-term approach means your running costs can be much lower."

Zinnia, a primary school teacher, masterminded the interior, which is painted and colourwashed with Keim, a hard-wearing mineral paint, and bought a beech tree from a local tree surgeon to provide the wood for the stairs and windowsills.

The Wilkinsons believe the evidence proves that eco-friendly features are cheaper as well as kinder to the planet. "We bought a solar panel. I was a bit wary of having it on the roof and doubted it would really be efficient in the British climate," admits Kester.

His scepticism was overturned, however, when the family discovered the panel generated enough heat to keep their water at 36C even under the leaden January skies, and reduce hot water bills by about 10%. They spend only about £240 a year on hot water and heating for their 2,152 square foot family home.

"What puzzles me is why developers don't take notice of the experience of people like us. They don't even bother putting in improved insulation. That alone would improve a property's eco-rating and save money for the owners, as we've proven", says Kester.

Even with this well above-spec property, its valuation when completed was well in excess of the total spent by the Wilkinsons.

Case study: Jez and Rachel Lavers

For Jez and Rachel Lavers, building their five-bedroom, three-storey dream home in Plymouth was a family affair that lasted four years. Jez and his brother Chris bought an old cottage on a plot big enough for two homes near where their parents self-built a home back in 1972.

"All the family were roped into reclaiming bricks, stone, slate and other items either for sale—to go into the building fund—or to use in the new construction" says Jez. During the build he and Rachel had full-time jobs and lived initially in "a damp little caravan" on site and then in a small attic flat; even so, they did most of the work themselves and only used professionals for external rendering and fixing roof slates.

"Chris and his wife Zoe were busy on their house next door. My parents committed huge amounts of labour, know-how, previous experience and support" he says.

To cap it all Rachel gave birth to baby John, just before the family finally moved in back in the autumn of 2008. As a bonus the house—which has individual room insulation, zoned heating controls and a thermal store—was in 2009 awarded a Norwich & Peterborough Building Society prize for best self-built eco-home in the UK.

Resource centre

Useful websites about all aspects of self build include: *www.selfbuildabc.co.uk*; *www.self-build.co.uk*; *www.buildstore.co.uk*; *www.ebuild.co.uk*; *www.selfbuildland.co.uk*; *www.homebuilding.co.uk*; *www.newbuildtimberframe.co.uk*; *www.potton.co.uk*; *www.selfbuildit.co.uk*; *www.selfbuildinnovations.com*; *www.selfbuild.co.uk*; *www.thehouseplanner. co.uk*; *www.segalselfbuild.co.u*k; *www.communityselfbuildagency.org.uk*.

If you want to go on a course there are free taster sessions at the many self-build shows that are held around the country (*www.homebuildingshow.co.uk* gives details of the largest shows, although there are many others held several times each year). The website *www.selfbuildcourses.co.uk* lists two-day courses starting at about £300, ranging from topics such as locating plots, winning planning consent, design, project management and budgeting.

In the Appendices of this book you will find details of how to deal with estate agency problems and information on Home Information Packs for use when you sell your completed home.

Redeveloping

Introduction

Buying a home to redevelop is one of the most obvious and long established ways to use bricks and mortar to earn you money. It can be both a lengthy and complicated process, involving substantial outlay before you see a return—but, if the redevelopment work is of a good quality and pitched at the right level, and if the timing is good, it can be a substantial earner.

First decision—are you redeveloping a home to let, or redeveloping to sell?

The answer to this question will determine where you buy and what you buy.

A property you intend to *redevelop and then let out to tenants* for a few years while prices rise and which you may then sell at a good profit (hopefully) will have to be in one of the suitable renting locations. Likewise, it will have to be a type of property that is likely to attract the sort of tenants who look to the private rented sector for a home.

Of course, in theory any type of property will find a private tenant (large families rent substantial detached houses in some areas, for example). However, analysis of the private rented sector shows very clearly that the substantial majority of tenants are under the age of 35—single young professionals living solo or with a friend, or students again living singly or in groups, and frustrated would-be first-time buyers who have to rent because they cannot afford to buy.

So, you might think, the logical move is to redevelop a property—say a two bedroom apartment, or a terraced house that can be converted into two apartments—in an area where there are workplaces and social facilities attracting large numbers of under-35s. But you absolutely must consider the existing supply of homes in these areas.

Most big city centres already have far too many flats, and this was one of the problems that directly led to the much bigger price falls experienced in this sector than elsewhere during the 2007 housing market downturn.

For example, estate agency Knight Frank says 87% of homes in central Leeds are one- and two-bedroom flats, with 40% privately rented. Some 70% of new homes in central Manchester since 2001 have sold to landlords. In Sheffield, the figure is 65% since 2002, and there are similar statistics and trends from Liverpool and Newcastle.

This is by no means to say that it is impossible to find good city centre locations where redeveloping a flat to rent out might not be profitable, but it is clear that the centres of the largest cities are saturated —this kind of market information (a lot more of which is in the Buy-to-Let chapter) must inform your thinking before you decide what sort of property and what location to select for your redevelopment project.

If you intend to redevelop to sell, pretty much immediately upon completion, you have a different set of criteria. Crucially, you must judge *who would buy your completed, redeveloped property*. This will help you work out the type of property and area where you should buy a wreck.

The following are typical potential buyers:

- **Families**, who will typically have three or four people in a home, possibly all adults, meaning you will have to ensure there are enough bedrooms, bathrooms, a decent sized kitchen, storage areas, a garden of some kind, and even garaging in some cases.

- **First-time buyers**, who will typically be looking for a property with facilities that lend themselves to busy, young individuals—so they will need one or two bedrooms, a reception area, probably a small kitchen as they may eat out or at work, but they may not require a garden and probably won't want the commitment of maintaining one.

- **Retirees**, who will typically be downsizing from a large house they can no longer manage. They will want a compact home with easy-to-access facilities, fixtures and fittings in the right place and at the right height, and very low maintenance. They probably will want a garden or outdoor space of some kind but will not be able to maintain them—so the simpler they are, the better.

- **Students**, who will typically want individual bedrooms for each person and perhaps only one modestly sized reception room, a compact kitchen and ideally no outdoor space—but plenty of storage, including for bikes, and good electrical wiring and even wi-fi throughout the property.

- **Holiday home buyers**, who will typically buy in areas which have not only outstanding beauty but also good restaurants, relatively easy road access, and plenty of storage (especially if they are in a seaside area) and even more storage space in case they want to keep their own possessions but rent out the property to holidaymakers.

Second decision—what type of property?

Although some amateur developers will be property enthusiasts and will be able to differentiate a Sunshine 1930s house from an Arts and Craft home, many others will not. It would take a book in itself to cover every possible period and type of building, but every developer should at least know these definitions:

- **Apartment**: this is the modern name for "flat" and is a self-contained home, commonly with just one or two bedrooms but in theory with an unlimited number of bedrooms, and the unit is usually one of many in a larger building. So for example, a house converted into apartments may typically have three units—but an apartment in a purpose-built block may be one of literally scores or even hundreds. Most apartments are "lateral" stretching across one floor of the building, typically (but not always) with little outdoor space.

- **Maisonette**: this is an apartment that exists on two storeys of a building, with an internal staircase linking the storeys.

- **Duplex**: this is a vague term but is most commonly applied to purpose-built apartments which can operate either as single units or which can be linked. So in theory, two adjoining two-bedroom apartments can be easily modified to form one four-bedroom apartment.

- **House**: this is a single home, usually built on two or more floors, designed to function as a unit for one household, typically consisting of two adults and some children.

- **Studio**: this is the modern term for "bedsit" and it designates a small apartment. The term is typically applied to the smallest apartments in a block. Usually studios are designed for only one resident, and sometimes the main room operates as a living area and a bedroom, with the bed folding away outside of sleeping hours.

- **Loft**: this is, technically, the space in the roof of a house or block, but "loft-style" apartments are regarded as large, spacious apartments often in converted industrial buildings and with fashionable "New York loft" characteristics such as exposed beams and pipes, wooden flooring, minimal fittings and large open spaces.

Of course, you can buy one type of property and, during its redevelopment, convert it into one or more other types—so, for example, you buy a house and convert it into one studio apartment and two "conventional" apartments. Your choice will be determined by the original building, the views of planners, your budget and building skills, your target renter or buyer, and the appropriateness of your plans for the area.

You must *avoid over-developing a property*. Even if you get planning permission to extend as part of your redevelopment project, you should be wary of turning, say, a two bedroom house into a five-bedroom house in an area where most of the nearby properties are small. It is often the case that streets are "known" by estate agents and residents as being typically full of "small" or "large" properties and these set the price benchmarks.

So if you redevelop and turn your project into a five bedroom house which you wish to sell at £500,000, it will appear very expensive indeed if the surrounding homes are all £200,000. So remember, horses for courses.

Calculate your budget

Having calculated where you should buy, what sort of property and for what purpose—including the eventual buyer whose purchase will provide your profit—you must then calculate your budget. This is not a detailed proposal (which will have to wait until you see the precise scale of work required by the property you want to buy) but it is an *overall business plan*.

Doing it now will help you focus on what skills you have (as you will have to budget for buying-in those skills you do NOT have), and it will ensure you begin to assess your own financial capacity.

For example will you have to, and are you prepared to, mortgage the home you live in to fund the purchase of the wreck and the redevelopment? Or will you be able to afford to keep your home and its payments AND borrow to purchase a wreck AND borrow more to fund the building work?

Once you have your overall business plan worked out, you can judge how much money (if any) you need to borrow from a lender.

Applying for a mortgage

We all know mortgages are tougher to obtain than at any time in recent years and this applies to those wishing to redevelop as much as to first-time buyers or long-standing owner occupiers.

If you are very inexperienced, make sure you have these questions answered before you sign anything—the answers will help you, providing you have furnished the lender with accurate information about your income, proposal and outline costings:

- How much can I afford to borrow?
- How can I tell which mortgage rate is best?
- What is the best type of mortgage for me?
- How should I repay it?
- Can I make lump sum payments?
- Are there any redemption penalties?
- Does this mortgage come with insurance?
- What other charges will I have to pay?
- What happens if I can't pay?

Most people embarking on a redevelopment project will already have had mortgages, either now or in the past. Therefore, using all the information they have about the project they wish to embark on, you should choose from:

- a **repayment mortgage**, which means every payment you make pays off some of the interest on your loan and some of the "capital" debt (that is, the sum you borrowed to begin with)
- an **interest-only mortgage**, which means your payments are only for the interest, while the original "capital" amount owned remains the same. Many redevelopers will choose this mortgage because they believe the money they achieve when they sell will cover the capital and still leave a profit.

When you pay the interest (either as part of a repayment mortgage, or the interest-only option), you can choose between mortgage products with:

- **variable rates**, which mean that as the Bank of England base rate changes so the interest rate on your mortgage will change
- **fixed rates**, which mean the interest component of your payments stays the same for a fixed period—whether the Bank of England rate rises or falls during that time
- **discounted rates**, which mean you pay a sum which includes a discount off the lender's variable rate, for an agreed term.

Buying your wreck

Remember that every estate agent's contract with a client emphasises that he or she is charged with getting the maximum amount of money for the property on sale. Just because a home is a wreck does not vary that emphasis, so to maximise your profit you need to undertake a thorough inspection of the property and you need to barter down the price wherever possible. Look for the following points:

- **Front garden and entrance**: check the quality of drives, paths and railings; insist that outbuildings are cleared and clutter removed by the seller (as it will save you some money and time doing it).

- **Reception rooms**: it sounds obvious, but check that shutters and radiators work and ensure the place is in good decorative order— if it isn't, offer less, even though you are going to rip the place apart. If there are DIY jobs unfinished, you should also offer less. In particular, check for cracks and decorative problems and ask if there are rewiring certificates available.

- **Kitchen**: check plumbing and ask for servicing documents on boilers and equipment; ensure cupboard carcasses are sound; check there are enough power points and storage spaces.

- **Bathroom**: look for poor seals on baths and WCs and any sign of dampness; check plumbing and the quality of suites. Are there already enough bathrooms for your project's needs? Is there enough hot water for your target buyer?

- **Bedrooms**: measure to see if bedrooms are large enough for beds, wardrobes and drawers. This will help you work out if you need to move walls or even substantially reconfigure the entire building.

- **Attics and lofts**: check size, condition, whether it has easy access, good lighting and adequate insulation—and make sure you go up there in person.

- **Flooring**: take up carpets and check the soundness of floorboards underneath.

- **Grounds**: check that boundary fencing is adequate, outbuildings are sound and rubbish cleared before completion.

- **Exterior**: check windowsills and doorframes for signs of rotting; ensure gutters and drains are in good condition and easily accessible.

Is it worth it?

How much profit you make depends very largely on how good a deal you can make when you buy the wreck to begin with. Don't just go along with the estate agent's patter about the number of bedrooms— you may be about to change the layout anyway. Instead, concentrate on how much space you get.

Increasingly, property gurus believe buyers should abandon the uniquely British way of judging the size of flats and houses by their number of bedrooms. Instead, buyers should take the more statistically rigorous approach of measuring floor space.

Although this technique (known as "pounds-per-square-foot" or "£/psf" in the trade) is not helpful when comparing unusual properties, it lends itself to most purchases of standard homes.

So if, say, you view three semi-detached houses in one area, ask the estate agents to state the "net internal floor space" so you can compare them exactly. You then simply divide the number of square feet by the price and get the £/psf for each.

So, for example, in Bury St Edmunds in 2008 there were two two-bedroom flats for sale in a converted listed house called Ashfield Grange. One was priced at £227,500 and appeared better value than the other, priced at £275,000. However, the former had 890 square foot so was £256psf, while the latter had 1,225 square foot, working out at £224psf.

It is hard to gauge how big or small one property is compared to others, so working out the net internal floor area and applying a cost-per-square-foot gives a more accurate basis for valuations, rather than gut feel or guesswork.

Historically, many estate agents and developers have been reluctant to specify square footage, especially for apartments and conversions of older, larger houses. Concentrating on rooms, not floor space, means few buyers know that a typical new-build home today is actually 55% smaller than an average one built back in 1920. And although house sizes in modern homes have not changed in the past 20 years, they typically have 20% more rooms today as home offices and en-suites have proliferated, while bedrooms have reduced in size.

The so-called Parker Morris Standards, introduced for council housing in 1967, say there should be 355 square foot of internal space for the first occupant of a property, with each additional resident getting another 140 square foot. So just over 40 years ago, a one-bed council flat for two people would have had at least 495 square foot.

Compare that to today. Research for this book revealed that in 20 Barratt Homes schemes across England, the majority of one-bedroom flats were 5% to 15% smaller than 495 square foot—although some were larger. Even less floor space was found in the studio flat unit favoured by Barratt and many other developers, where sizes routinely go as low as 360 square foot, or under 75% of the recommended minimum for two occupants four decades ago.

Many experts say a switch to pricing by floor-space would allow buyers at all levels of the market to see beyond the "bedroom number" marketing of house builders.

But there are shortcomings to this method. Unique, one-off homes do not lend themselves to floor space comparisons. In addition, make sure your comparative square foot measurements are like-for-like—if you get two different measurements on two different properties, ensure they both include (or exclude) roof terraces, hallways, wine cellars, staircases, storage facilities and the like. Bear the following points in mind:

- The smaller the property, the higher the cost per square foot.
- Ground rent and service charges will affect the £/psf.
- Also calculate the price per cubic foot—high ceilings push up values.
- In new schemes, £/psf is higher on top floors because of the views.
- Location can affect £/psf, but condition and interior rarely have any effect.
- The property's lease length must be considered in the case of leasehold flats.

Auctions

Becoming an amateur developer is ambitious but auctions are the right places to find homes in need to renovation.

Sellers are often advised to choose auctions over estate agents if they need a quick sale, or because a property has problems and can sell only at below-market price, or because a property is ugly (like a flat above a shop) and will not sell easily in the beauty parade of homes in an estate agent's window.

Auctions do occasionally generate fevered bidding but most are surprisingly calm affairs with up to 150 properties being offered over three hours. But because the sale lots have unusual characteristics and sometimes limited scope to visit, buyers must be cautious and research is essential.

Common problems include the need for re-modernisation, the existence of sitting tenants in a buy-to-let flat, or structural and subsidence difficulties that mean a buyer will not be able to obtain a conventional mortgage and must risk a cash purchase.

Many auctioned flats also have short leases that must be extended by solicitors unless a new owner accepts a reduced re-sale value in the future. Some auction properties may be in poor locations such as near anti-social industrial sites, on the edge of a busy road or next to a site with permission for major development.

But auctions are perfect for finding homes needing renovation. Remember that your re-sale markets will be different—owner-occupier or investors in London, and perhaps second home owners or holiday lettings investors in the South West. Your refurbishment budgets may therefore differ from place to place.

Once you have researched a property coming to auction, the only obstacle is the bid itself. Psychology is important here.

Experts advise you to arrive early, sit where you can see what's going on and who you are bidding against. If the bidding is fairly slow, don't show your interest too early. Make sure the bids work out so that it's you who is bidding when the price reaches your upper limit. This requires a certain amount of mental agility. But don't get carried away and exceed the limit you have set.

If you reach your personal upper limit but then the property is withdrawn because it has not met the seller's reserve, not all is lost. A private deal can often be struck after the bidding has concluded, with the auctioneer acting as middle-man while you and the seller negotiate. You will have to make a quick approach to an auction employee working near the auctioneer to get this arranged.

Creating your team

You will need to identify and recruit a number of skills to work on your project. The exact skills, man hours and plant required will depend upon the scale and nature of your redevelopment, but they are likely to include the following:

1. **An architect**—this will be almost always necessary. Find one through the Royal Institute of British Architects (RIBA) (Tel: 020 7580 5533, www.architecture.com) which will have a list of those in your area, with experience of this kind of redevelopment work. Set up an initial meeting to discuss your project. This may be free, even if you do not commission the architect, or it may involve a small fee. If you wish to pay, the architect will then project manage for you but if you want to be more hands-on you can find other key personnel yourself.

2. **A project manager**—as with the self-build projects set out in Chapter two, a project manager can be a valuable member of the team. To find one, scour websites on the internet or contact the Project Management Institute (*www.pmi.org.uk*) which will guide you to experienced people in your locality.

3. **Contractors**—these are likely to involve builders, plumbers, electricians and, depending on your project, occasional work from specialists like landscapers. Which?, previously the Consumers' Association, advises that you get these details:

    (a) itemised cost of materials, and the type used
    (b) likely time expected, in days and hours
    (c) a fixed cost per day or for the work.

Which? also recommends that you organise contracts. For the smallest jobs, say under £5,000 in total, these could be simple "letters of agreement" setting out:

*   a description of the work to be done
*   confirmation of who does what
*   price for the work and who is going to do it
*   confirmation of the building regulations affected
*   how and when the work will be paid for.

More formal contracts are available from the Federation of Master Builders and Trading Standards departments (see the Appendices at the end of this book) and can be drawn up or at least overseen by solicitors. These will be required for your main builders who will be undertaking the long-term work. These contracts should include:

*   the start date and finish date for work
*   what happens if the contractor cannot complete the work
*   how the contractor will comply with statutory requirements such as building regulations
*   how any damage will be made good
*   how sub-standard work is dealt with
*   clearance of rubbish
*   work guarantees
*   how and when payment is made.

Applying for planning permission

Although every area has its own local authority, including councillors sitting on planning committees and professional, skilled planning officers to advise them, the guidelines they follow are determined nationally, at government level. There are relatively minor differences between England, Scotland, Wales and Northern Ireland but there are many common features.

So, for example:

- Planning applications are decided in line with the local development plans drawn up by local councils—so you can quite easily check that you are not planning something inappropriate, like converting a house into flats in an area where this is already designated as undesirable.
- Planners will want to know the size, layout, siting and external appearance of buildings you construct or modify.
- The proposed means of access, landscaping and the impact on the neighbourhood.
- The appropriateness of the infrastructure, such as roads and water supply, for the proposal you make.

Government guidelines say it is not necessary to make the application yourself so you can appoint an agent (for instance, an architect, a solicitor, or a builder) to make it for you.

Architects and local council planners will help you on specific proposals (the latter will often give some assistance for free). No one should be under any illusions here—almost every redevelopment project requires substantial work and that means planning consent, but it is worth noting that many individual pieces of work do not require such permission these days. Here are some of the basic guidelines about the most common pieces of work done in redevelopment projects:

- **Basements**: Converting an existing residential cellar or basement into a living space is in most cases unlikely to require planning permission as long as it is not to turn the "storey" into a separate unit. Excavating to create a new basement which involves major works, a new separate unit of accommodation and/or alters the external appearance of the house, such as adding a light well, is likely to require planning permission.

- **Chimneys**: Fitting, altering or replacing an external flue, chimney, or soil and vent pipe is normally considered to be permitted development, not requiring planning consent, under certain conditions.

- **Conservatories**: A conservatory is considered as permitted development not requiring planning permission so long as it fulfils criteria and sizes outlined at *www.planningportal.gov.uk*.

- **Demolition**: In most cases you will not need to apply for planning permission to knock down your house or any of its outbuildings, unless the council has made an "Article 4 Direction" restricting the demolition or alterations you could normally carry out under permitted development rules. However, it does not automatically follow that you will get planning permission to build any replacement structure or to change the use of the site.

- **Drains and drainage**: Repairs and maintenance are not required but sometimes you need to apply for planning permission for some of these works because your council has made an "Article 4 Direction" withdrawing permitted development rights because of local circumstances (like old and fragile drains). So always check with the planning authority.

- **Driveways**: You will not require planning permission if a new driveway uses permeable or porous surfacing that allows water to drain through—gravel, permeable concrete block paving or porous asphalt is fine—or if the rainwater is directed to a lawn to drain away naturally. If the surface to be covered is more than five square metres planning permission will be needed for laying traditional, impermeable driveways that do not control rainwater running off onto roads.

- **Extensions**: If your redevelopment project is very small, it is possible you will need much less contact with planners than you might expect. Under new regulations that came into effect in 2008—formally called the Town and Country Planning (General Permitted Development) (Amendment) (No. 2)—an extension or addition to a home is considered to be 'permitted development', not requiring an application for planning permission, subject to limits and conditions set out at *www.planningportal.gov.uk*. Be realistic—it is very unlikely that your project will be so small that it will not need planning consent, but it is at least worth checking.

- **External walls**: You do not need to apply for planning permission for repairs, maintenance or minor improvements, such as painting your house. If you live in a listed building, you will need listed building consent for any significant works whether internal or external. If you live in a Conservation Area, a National Park, an Area of Outstanding Natural Beauty or the Broads, you will need to apply for planning permission before cladding the outside of your house with stone, artificial stone, pebble dash, render, timber, plastic or tiles.

- **Fences and walls**: Whether or not you need to apply for planning permission when building or replacing a fence, garden wall or gate depends on a number of factors including height and position.

- **Gardens**: Decking or other raised platforms will be regarded as permitted development providing it is no more than 30cm above the ground and so long as, when taken together with other extensions and outbuildings, the decking or platforms cover no more than 50% of the garden area.

- **Lighting**: Minor domestic light fittings are not subject to planning controls but if you plan external lighting for security or other purposes, you need to ensure the intensity and direction of light does not disturb others.

- **Loft conversions**: Planning consent is required where you extend or alter the roof space and it exceeds specified limits and conditions set out on *www.planningportal.gov.uk*.

- **Roof**: You do not normally need to apply for planning permission to re-roof your house or to insert roof lights or skylights. New rules introduced on 1 October 2008 allow for some roof alterations without the need for planning permission—for full details see *www.planningportal.gov.uk*.

- **Solar panels**: In many cases fixing solar panels to the roof of a single dwelling house is likely to be considered "permitted development" under planning law with no need to apply for planning permission.

- **Windows and doors**: Planning permission is not normally required for repairing, fitting or replacing doors and windows, including double glazing. But if your redevelopment project is on a house that is listed or is in a conservation area, you may need informal consultation with the planning authority.

The planning process

The formal process is laid out at *www.planningportal.gov.uk* which explains that points that will be looked at in special detail will be:

- the number, size, layout, siting and external appearance of buildings
- proposed means of access, landscaping and impact on the neighbourhood
- availability of infrastructure, such as roads and water supply
- the proposed use of the development.

It is fair to describe the process as bureaucratic, slow and clumsy—but it is unavoidable. The order of the process is this:

1. You or your architect or project manager contacts the local planning department for advice.
2. You then request an application form, or download one from the Planning Portal at *www.planningportal.gov.uk*.
3. You submit either (a) an outline application which deals with the major principles of the scheme with an agreement that the details will be looked at later, or (b) a full application listing all details.
4. You then submit the application with supporting documents, like the architect's drawings.
5. The local planners validate the application and may call for additional documents to help decision makers.
6. The council then publicises the application, writing to neighbours and interested parties, and seeking views within a short deadline, usually of a few weeks.

7. A planning officer or a committee of elected councillors then decide for or against the application—the applicants may be heard in public, and so may their opponents, if the issue goes to the committee.

8. The application is then agreed outright, agreed with conditions, or rejected.

9. If rejected, the applicant can change the plans and resubmit them, or can appeal against the rejection (the latter can take many months and usually is decided behind closed doors).

10. If agreed, the applicant should proceed with the work.

Case study: winning planning permission

Ben and Susie Green do not need convincing of how difficult it can be to obtain planning permission—they say they wasted nine months and thousands of pounds on unnecessary planning applications and bureaucracy.

The couple renovated an end-of-terrace Victorian house three years ago to create a small number of large rooms, the reverse of the typical property of that era, which has a large number of small ones.

"We dug out a basement and made a side extension and turned the loft into a room", explains Ben, an investment banker. But the idea became a bureaucratic nightmare when Wandsworth Council planners unexpectedly rejected his proposals.

"Each time we had something rejected it meant scrapping plans, more meetings with architects, new plans, lots of paperwork, new consultations and more delays. It took a tremendous amount of energy." The couple spent six months doing the real work, yet nine months and "several thousands of pounds" on abortive planning proposals.

"The roof was the main problem. We'd submitted a conservative design but that was rejected", he says. "Then we discussed loft conversions identical to other end-of-terrace houses on our junction but were told past precedent didn't mean our similar scheme would be approved. What was acceptable a few years ago may not be acceptable now, we were told. We didn't know what to do."

Eventually, the Greens hired an architect who found that by building a more flamboyant glass-roofed loft conversion to a lower height, they did not need planning consent at all.

The work

Your professional advisors and contracted builders—not to mention the planning watchdogs—will determine what work is required. It will be unique in every single case, but assuming your redevelopment is comprehensive throughout your property, there will be a traditional "staging" of activities. This involves:

* **preparation**, which takes in most of the stages outlined above, plus the acquisition of materials for the "first fix" stage below, and the construction of foundations for any extension for your redevelopment project

* **first fix**, which moves the project from foundation stage through the renovation or construction of floors, ceilings, internal walls and cavities for utilities (water, electricity, gas) to the plastering of the internal walls

* **second fix**, which involves the connection of electrical fixtures, fitting of internal doors and windows, and gives the property a "near complete" feel

* **completion**, which takes the property to its formal finish in such a state that an owner could acquire it within one week.

The duration of this work will, of course, depend on the scale of your exercise.

The Federation of Master Builders advises, quite reasonably, that much work can produce unexpected delays but, once fund-raising and planning has been completed (and this can take up to a year), it suggests these timescales for work:

* Two-storey extension: 12 weeks.
* Single-storey extension: 10 weeks.
* Additional loft: 2 weeks.
* Basement creation: 6 weeks.
* Garden/driveway paving: 1 week.
* Garage building: 2 weeks.

Nearing completion—put the home on sale

When you are within three months of completion, research the market via the obvious websites—*www.hbosplc.com* or *www.nationwide.co.uk* for prices and *www.hometrack.co.uk* for an indication of how many weeks and how many viewings it takes on average to sell a home in your area. Then get your property valued by an estate agent using this three-stage process:

- First, invite the correct agents. Different ones in your localities will have different specialities—some will sell mainly to first timers, others like family houses, some handle the top end, and so on. Obviously, get ones dealing in your kind of property. Ask how long they have had homes on sale and which have reduced prices because they overvalued them in the first place to win instructions. A pattern will appear and the best agents will emerge.

- Second, when you have about four good agents, resist automatically choosing the one suggesting the highest asking price. Ask for "comparables"—these are examples of similar homes sold nearby. Check how long agents took to sell them. Most of all, be realistic on prices and acknowledge that the market may not be buoyant, even by the time you complete your cherished redevelopment. Experts like Kate Faulkner, a property guru who has worked for developers, agents and runs her own consultancy, suggests you ask each short-listed agent for three prices—one would be an ideal asking price, the next is a price that would sell your home in six weeks and another that would sell it in two weeks.

- Third and finally, check if an agent can chase your sale. This is absolutely vital these days, because sales are still slow compared to the halcyon days of 2007. You should ask agents to demonstrate that they have a strategy to advertise your home on the internet, in the press, what client base they have looking for your sort of home, and so on. Negotiating a low agency fee may be a false economy, too, as it may drop the priority given to your home. Canny developers incentivise agents with a sliding scale of fees based on the sale price—the higher the price, the bigger the fee.

Case study: Jamie Campbell-Walker

If you think there is little in common between fast driving and slow housing markets, have a word with Jamie Campbell-Walker, a British world champion driver who regards developing houses to be the perfect complement to the race track.

"I race during the summer and work on houses in the winter. It's a good match" says Jamie, 35. He has developed five properties in the past, mostly small homes that he lived in during the work. But now he has spent well over a year on his biggest project, a vast seven-bedroom house in Wandsworth, south London.

"The house had been 12 bed-sits for Polish workers", he explains. "I bought it for a song in late 2006, received planning permission in May 2007 and got down to business."

He worked on it between race weekends and teamed up with Hertfordshire builder Mick Wadey to transform the property. "We ran 25% over budget because party wall disputes meant we could only develop one half of the house at a time", he admits.

The biggest task was turning a coal cellar into a 1,300 sq ft basement with 6ft ceilings. "It's a bit ambitious for a first full-scale development but it was worth it", he says.

The project and Jamie himself faced disaster in April when he broke his back in a 147mph crash at Monza in Italy. He escaped paralysis but spent months in a back brace. The smash wiped out much of his 2008 racing season and stopped him working on the house.

"It was a critical time but I recovered, although obviously the housing market is now in poor shape", he says. But he is undeterred and, banking on the market bouncing back in 2010 or 2011, will buy another house for a second long-term project.

His tips for developers in a downturn are:

- **Buy a wreck now**: "But only if your project will take 18 months or more, thus ensuring it won't be ready to sell until the market comes out of recession."

- **Check with planners first**: "Meet them, chat to them, get advice and don't waste money by finalising plans and hiring builders before you get consent."

- **Research your market**: "If you're building in a family street near a school, then target that audience and don't try to create a different market."

- **Assemble a team with the right skills**: "Nothing beats getting a builder involved from the start, even if it means you split the profit with him."

Listed buildings

You would be strongly advised to steer clear of listed properties to be converted, at least until you have a string of successful projects under your belt. Although the rewards may ultimately be high, because of the desirability of most listed houses, getting to completion stage will almost certainly be a nightmare...and an expensive one, at that.

There are three grades of listed building in England and Wales:

- Grade I are of exceptional interest.
- Grade II* are particularly important buildings of more than special interest.
- Grade II are of special interest, warranting every effort to preserve them.

There is a common misconception that listing only covers the outside or the main façade, but in fact the whole structure is included. All buildings built before 1700 which survive in original or near-original condition are automatically listed, as are most built between 1700 and 1840. After that date the criteria become tighter with time, so that post-1945 buildings have to be exceptionally important to be listed.

The vast majority of Britain's 500,000 listed buildings are houses, categorised as Grade II and described as "of special interest". All listed buildings are protected and any owner who wants to make changes to them must win Listed Building Consent (LBC). Local councils can give consent for some changes to Grade II homes but must get agreement from English Heritage for changes to higher-graded properties.

Some changes will also require "normal" planning permission on top of LBC, and in London there are further checks by some borough councils and the London Assembly.

But the number of listed buildings and the vigour with which councils monitor them have increased, just as personal wealth and the passion for makeovers have risen, too. The result? Some people have fallen out of love with listed buildings.

Eco features

It's difficult to be politically correct when saying this, but one point must be made—it is rarely the case that the cost of state-of-the-art eco features can be recovered in the sale price of a redeveloped home.

Constructing, maintaining and living in our homes accounts for 28% of UK carbon dioxide emissions, says the Government, so the issue cannot be ignored.

The average "old" house loses about 35% of its heat through the walls, 25% through the roof, 25% through draughty doors and windows and 15% through the floor—so as you will undoubtedly have improved those figures during your redevelopment, you should rightly boast about doing so when advertising the finished product. Also the so-called "Building Regulation L", already in force, sets out minimum standards for double-glazing, boiler types, and insulation levels in walls, floors and roofs which you will have met.

But there is no agreed definition on what package of features beyond these will make a home more or less eco-friendly. Flats are neither better nor worse than houses, but bungalows are considered by some to be "space-greedy", as they are so low-density.

Experts, such as the Town and Country Planning Association, say sustainable homes should: be built with materials selected from natural, renewable or recycled sources; have energy-efficient designs and preferably be south-facing to harvest heat from the sun; include water-saving appliances and water-recycling systems; accommodate recycling bins; and by their location and design, reduce car dependency or, across whole developments, feature innovations such as car-pools.

You can get more details from the Town and Country Planning Association's website at *www.tcpa.org.uk* but remember that if you want to maximise your income, you may have to compromise.

Resource centre

In addition to the websites listed above, check out these craft sites:

You can still check the Council for Registered Gas Installers (CORGI) at *www.corgi-gas.com*, but from summer 2009 the new name for CORGI is "Gas Safe", see *www.gassaferegister.co.uk*; the Electrical Contractors Association at *www.eca.co.uk*; the Federation of Master Builders at *www.fmb.org.uk*; the Institute of Plumbing at *www.plumbers.org.uk*; the National Association of Plumbing, Heating and Mechanical Services at *www.apha.co.uk*; the National Federation of Roofing Contractors at *www.nfrc.co.uk*; the Painting and Decorating Association at *www.paintingdecoratingassociation.co.uk*; and the Heating Ventilation Contractors' Association at *www.hvca.org.uk*.

In the Appendices of this book you will find a draft contract to prepare for those working on your redevelopment project, details of how to deal with estate agency problems when your property is sold, and information on Home Information Packs which you will encounter when you sell your completed property.

Buy-to-Let

Introduction

Buy-to-let is a great way to provide a public service by creating good quality accommodation for renters, as well as making an income for the landlord investor.

Its principles are straightforward and work well for investors so long as they make their choices wisely, from the location, type and price of the property when they purchase it, through the rental charge and on to the timing of when they sell.

Yet it has become, for some, a dirty word. A few blame buy-to-let for the slide in the British housing market, because, they claim, too many landlords paid too much for small homes which consequently priced-out first-time buyers.

That is nonsense—Germany, for example, has had a buy-to-let sector that has been much larger than Britain's, and its house prices have hardly risen in 20 years. In reality, buy-to-let is the product of a host of changing social forces in Britain.

The Housing Act 1988 de-regulated the rental market and provided safeguards for both tenants and landlords at roughly the same time that new working practices and growing personal aspirations amongst school-leavers led to much more mobility—including people moving across national boundaries, as well as around Britain.

That is the trend that led to what we now call buy-to-let and it shows no sign of changing, even in the current recession. Buy-to-let can still be a profitable exercise for individual landlords and the main reason behind this is simple—Britain needs a private rented sector and demand is growing.

Savills, the estate agent, says that the long-term trend in the UK is to reduce the country's fetish for home ownership, while renting increases in popularity.

It is a gradual move—71% of all households were owner-occupied in 2003 and that dipped only slightly to 68% in 2008—but it is a pattern that will almost certainly accelerate in a recession. What is more, only 37% of all households are buying with a mortgage, which means that 46% of owner occupied households own their homes outright.

This indicates, Savills says, that owner-occupation is dominated by those with substantial equity in their property rather than borrowing from scratch. Much of this change has been down to the emergence of a much larger, and better quality, private rented sector in the UK.

This sub-market has been expanding at an unprecedented rate since 1988, growing by an annual 3% until the credit crunch hit. By 2008, 14% of all English homes were privately rented compared with 7% in 1989. If that continues, there will be a long-term shortage of property to rent.

In addition, separate research by Lloyds TSB insurance shows that the number of single households will increase by almost two million by the year 2019, driven by the desire of young adults to live alone and the increasing affordability of solo living after the price falls of 2008 and 2009.

In the City of London and Westminster more than half of homes have just one resident now. By 2019, it is predicted that two in every three homes in the City will be single person dwellings. Other locations identified for solo living include Corby in Northamptonshire, North Cornwall, Basingstoke in Hampshire, Basildon in Essex and Stevenage in Hertfordshire.

Yet more research, this time from the Government's Office of National Statistics, which predicts four in 10 homes will be occupied by someone living alone by 2030. The number of single-person households will rise from 6.8 million in 2006 to 10.9 million in 2030. It means that around one in five of the total population will be living alone in England, up from 13% in 2006.

Building levels are not likely to keep anywhere close to this demographic surge, so renting is likely to be the answer for a lot of singletons. Suddenly, buy-to-let doesn't look such a bad bet after all, does it?

There are 10 golden rules of buy-to-let which all new investors must follow, and these are set out below. They do not guarantee big returns, but they make them much more likely. But before we move to them, here are some big questions to consider.

Question 1: Do you want buy-to-let as an investment or a pension?

There is a difference. An investment is something you will draw upon to maximise profits. If the timing is wrong, as we saw in the recessions of the early 1990s and again in 2007-2009, investors who wanted the largest "profit" on the sale of their buy-to-let properties simply held on to them until the market bounced back.

Those using buy-to-let as a pension and wanting to sell their property upon reaching their retirement age, may find that it is NOT the time to sell if the market is falling. Yet they may feel they do not want to wait—so could sell in order to get at least some money, yet ultimately end up even making a loss on the project.

A way to avoid this is either to get into buy-to-let at an early age— 30s or 40s, for example, which gives between 10 and 20 years at least before retirement, and plenty of time to choose when to sell without the imperative of a looming deadline. Then sell at the top of the market, and invest the money in a safer (if less interesting and less well rewarded) area.

Question 2: Do you want rental yield or capital appreciation from buy-to-let?

Some quick definitions first.

Rental yield is the rental income received each year, expressed as a percentage of the purchase price. So if your flat costs you £200,000 to buy (including purchase costs like legal fees and stamp duty) and you receive £1,000 a month rent, then you get £12,000 in a year or 6% yield (because £12,000 is 6% of £200,000).

Capital appreciation is the "profit" you make, again after expenses have been taken into account. So if you bought that flat for £200,000 and sold it for £280,000 some 10 years later, the capital appreciation is 40% (because the additional £80,000 is 40% of £200,000).

Ideally, of course, you will want a rental yield that gives you a continual income AND some capital appreciation when you eventually sell. How can we achieve this?

The best way to ensure the annual yield provides some profit is to ensure that you buy a property where your monthly expenses can be as near as possible guaranteed to be lower than your rental income.

Buy-to-let mortgage lenders demand you have at least a 15% deposit, sometimes as much as 25%. The rental income will have to be 125% to 130% of the monthly mortgage payments to give you a vital "cushion" if there are short periods where you have an empty property when tenants change. This cushion also helps you cover maintenance costs, occasional decoration and ground rent. So if your mortgage will cost you £500 a month, your expected rental income must be £650—and lenders will get this professionally valued before you receive a penny.

However, mortgage payments on even large borrowings need not be huge because it is essential on a buy-to-let to get an interest-only mortgage (that is, not repaying the main loan—just the interest on it each month). This is because it makes monthly payments much lower, and also because as a buy-to-let landlord you receive tax relief on the interest element of a mortgage—officially, recognition that you are providing a valuable service.

Even on a successful buy-to-let project, income from rent after expenses have been taken into account may not be vast—say, £100 per week from each property. If you have only one property, that income will only be £5,200 per year.

So many investors also look to capital appreciation in the long-term to earn them a healthy reward for their investment. This is where, in a recession, the canny buyer can do well.

If properties are, say, 40% down in price from their 2007 "highs" then an opportunity exists to buy at the newer low prices and wait—and hope—for prices to rise again.

Of course, as we all know, they may not rise to their past levels and if they do rise they may fall again as house price cycles take their course—so there are no guarantees. However, after previous severe price falls prices have risen again, and there is plenty of evidence to suggest this will happen again.

So how long will this take?

Estate agency Knight Frank indicates that prices will rise again in most regions of the UK from 2010 onwards, very slowly taking as long as six years to regain their 2007 highs. This is, of course, merely on forecast and there are several others—this is probably a "middle of the road" one, and there are many others that suggest faster rises, and others still that suggest even slower rises.

But interestingly, most commentators agree that prices WILL rise—so profits are there to be made by those who get in early.

Region	By late 2009	Rises begin	Back to 2007 highs?
Greater London	–26%	2010	2016
Central London	–29%	2010	2016
South east	–23%	2010	2015
South west	–21%	2010	2015
Yorkshire	–23%	2010	2015
East Anglia	–27%	2010	2017
West Midlands	–22%	2010	2017
East Midlands	–21%	2010	2016
North West	–22%	2010	2016
North of England	–18%	2010	2015
Scotland	–15%	2011	2015
Wales	–22%	2010	2015
Northern Ireland	–44%	2010	2020 or later

Source: Savills estate agency, 2009

I repeat that this only a forecast, but it is one which shows the wisdom of looking on buy-to-let primarily as a long-term investment of 10 years or more, not depending on it maximising profit at a certain date to coincide with, say, retirement, and it shows that capital appreciation can be earned from 2010 onwards if the prediction is correct.

Having considered these factors, let us now move on to the *10 golden rules of buy-to-let*.

The 10 golden rules for buy-to-let

Rule 1: Choose your sector

These days there are several long-term renting sectors.

Long-term **"corporate" lets** to executives with company expenses are one type—evidently these tend to exist only where there are substantial business communities in south east England and close to other large city centres. Often, executives will be relocated with their families so the typical property for this sector is the well-appointed large family home with executive must-haves, which include the following:

- Modern equipment throughout.
- Large garage or area for parking.
- Privacy.

- Room for a member of staff, if required.
- Fashionable elements like a cinema room, inside pool or sauna.

Then, at the other end of the social scale, there is low-cost housing for **key workers**. These tend to be in public sector jobs like the NHS or local councils (in many areas teachers are now amongst the best paid and are not officially regarded as key workers for social housing purposes). The characteristics of this sector are as follows:

- Some tenants will want houses, but most will want apartments.
- They will be functional, fit for purpose homes, usually modern.
- They will be near transport hubs, local shops and major centres of work.

Similar properties are likely to be required by **frustrated first-time buyers** who rent while they save for their first home purchase.

Increasingly popular is the **student accommodation** sector.

Total student numbers in the UK have grown from 1.8 million in 1997 to 2.5 million in 2007, an annual growth of 3.3%. Knight Frank, the estate agent, says in recent years London's student population has grown even more at 4% per annum and that between 2001 and 2009 there was a particularly large 40% rise in the number of overseas students in the capital. Despite this growing demand, supply has nowhere near kept pace and the situation has deteriorated for students because the recession means more people outside of education are renting too—reducing the stock of homes available to those at college.

We talk at length about student property later in the chapter.

Rule 2: Choose your optimum size and type of property

This will vary according to area, availability and suitability for your sector. There is no point buying a bungalow for student letting when they would be more appropriate in a large house, nor using a run-down right-to-buy ex-council flat for letting to executives. There are national trends to consider, too.

One often forgotten fact is that while **former council flats and houses** now in the private market thanks to right-to-buy are inexpensive, they are often in areas considered unattractive to private sector tenants. This is a shame because the room sizes and general condition of the majority of council flats and houses are better than in their older private counterparts—but the stigma remains for some tenants, and for that reason a former local authority home is usually hard to let privately.

The same goes for **flats above shops**. In many ways this is a pity as the "flat over" has been seen as a cheap choice and safe haven for people moving to big cities since it first became common in Victorian times. It was traditionally large (the first ones were for the shop-owners themselves), always at the heart of communities, secure and near public transport.

Lenders are very wary of giving buy-to-let mortgages for above-shop premises. When the market turns sour, flats above shops are less desirable. Lenders worry about a shop which may be unobtrusive now but may change use to produce smells, noise or late night opening. There are three categories of shop assessed by lenders. Category A1 covers general stores like newsagents and food shops; A2 are professional services like banks and estate agents; A3 are the really difficult ones for mortgage companies, like restaurants, takeaways and pubs.

Be very wary of buying buy-to-let apartments in over-supplied city centres. By 2008 there were enormous over-supplies of buy-to-let flats (some sold but unoccupied, many not even able to find a landlord buyer) in Leeds, Sheffield, Liverpool, Manchester, Birmingham, Newcastle, Bristol, central London and Cardiff. This situation will change, slowly, over time so always research carefully—and don't take at first hand the comments or information provided by estate agents desperately keen to offload this glut of units. Even if you buy one at a bargain price, you may be unable to let it if there are literally hundreds or even thousands of near-identical units for tenants to choose from.

Also beware of buying over-specified properties, like those with concierges and heavily landscaped grounds. They look good and sound even better, but these services are not required by 99.9% of tenants and add vastly to landlords' costs.

Rule 3: Location, location, location—consider buying further afield

Research suggests that most buy-to-let landlords buy in their home town, but that does not guarantee the best rental income or capital appreciation. The market varies over time and some locations become popular because of regeneration projects, or expanding universities, or one-off schemes like a new airport. Obviously, if you manage the property yourself it must be near to your own home, but otherwise it could be anywhere.

Here are some simple tips about what to buy where:

- Tenants may not have cars so check public transport links.
- Some streets have reputations as "good" places to rent—try these.
- View at different times and days to see parking, busyness and neighbours.
- If you are letting to relocated executives, buy near good schools.

Rule 4: Work out your rental income

Before you buy you must calculate the likely income to ensure your mortgage and running costs are covered. Most letting agents advise their landlords to calculate their net yield (that is the total income expected over a year, less the total running costs, as a percentage of the capital value of the property). Tenants have to pay council tax and utility charges, but landlords normally pay the communal service charges if the property is in a block or on an estate with collective overheads.

Do not forget you may need to refresh the decor of your property between lettings and factor in an empty period each year either for gaps between tenants or to allow for updating.

The table below shows *average UK annual rates of return* on buy-to-let investment properties in early 2009, produced by the Association of Residential Lettings Agents (ARLA).

Gross annual yield is the percentage of the purchase price of the property recovered in a year from rent; voids per year are the number of days on average that a rental property is empty "between" tenants; and the five-year rate of return covers the proportion of the purchase price recovered in combined rental and capital growth in the five years to early 2009.

Region	Gross annual yield	Typical voids/year	5-year rate of return
Prime Central London	4.84%	28 days	10.56%
Greater London	5.13%	21 days	10.82%
South east England	4.93%	26 days	10.64%
South west England	4.46%	26 days	10.32%

Midlands	4.63%	26 days	10.43%
North west England	5.11%	31 days	10.70%
North east England	5.02%	26 days	10.70%
Scotland/Wales/NI	5.13%	31 days	10.72%
AVERAGE UK	4.87%	27 days	10.59%

Source: Association of Residential Lettings Agents

This table, also from ARLA, shows the differences in annual rates of rental return for houses and flats.

Region	Houses	Flats	Average
Prime Central London	10.40%	10.71%	10.56%
Greater London	10.73%	10.91%	10.82%
South east England	10.67%	10.62%	10.64%
South west England	10.25%	10.39%	10.32%
Midlands	10.61%	10.24%	10.43%
North west England	10.99%	10.41%	10.70%
North east England	10.82%	10.58%	10.70%
Scotland/Wales/NI	10.90%	10.54%	10.72%
AVERAGE UK	10.61%	10.58%	10.59%

Source: Association of Residential Lettings Agents

Rule 5: Calculate possible capital appreciation

The days of 15% or 20% price rises each year have gone, probably forever, so once the recession ends factor in only modest capital growth each year. But properties purchased in, say, late 2009 or in 2010 will probably benefit from several years of growth thereafter—if you regard buy-to-let as a 10-year-plus investment, you are more likely to see good returns.

Rule 6: Seek a competitive mortgage

These days there are websites that list subsets of the buy-to-let mortgage products that exist, and compare them for value for money. Expect to pay 0.75% to 1% more interest than on a comparable owner-occupier mortgage and always go for interest-only payments, as these optimise your tax status— that is, you can offset the interest payment against your tax bill. See *www.money-facts.co.uk* for more details.

Rule 7: Consider hiring a lettings agent

A lettings agent will charge you 15% to 20% of your rental income but offer excellent peace of mind. Without them, you will have to find and vet tenants, answer late-night queries from potentially cantankerous tenants, and find emergency plumbers when a pipe bursts just as you're starting a holiday. And remember that the agent's fee is tax deductable too. Insist that you see their records of lettings in the location and type of property you own.

ARLA is obviously pro-agent but has set out a very clear statement of what an agent does and why one is useful. Here it is:

> For the tenant, the agent has to find an instant home, usually at short notice, that will be pleasant to live in and in good working order. For the landlord, the agent is the guard dog, helping to protect an expensive asset by finding suitable tenants who will enjoy and care for the property, by managing the maintenance and repairs and by producing the income stream the owner requires.

To attract good covenant tenants the agent needs to be well-established in the area with a strong local reputation. It is this reputation—for professionalism, membership of ARLA with its requirements for qualified staff, client accounting and indemnity cover leading to fidelity bonding—that brings in the instructions from the property owners—and that is what attracts the tenants.

After a tenant's offer has been accepted, the letting agent is responsible for taking up references and making credit checks before preparing to change the utility accounts and drawing up the Tenancy Agreement.

It all begins with references, credit references from the bank or building society and personal references from the employer and, if applicable, from a previous landlord. These personal references are very important. There's a total stranger about to move into the property. Some clues to character are needed to match with the agent's own experienced assessment of the applicant. For the self-employed, references are expected from an accountant or solicitor.

Then, an inventory and condition report must be made. In this report, the entire contents of the property will be noted down, along with the condition of the carpets, curtains, furniture, wallpaper and paintwork. Both the tenant and the landlord, or their representatives should be present when this inventory report is compiled, and then they should sign each others' copies. It will be the basis for checking that the contents and the condition of the property at the end of the tenancy are the same as they were at the start.

The legal process is fair and equitable to both landlord and tenant, everyone will know where they stand.

Virtually every let nowadays is arranged as an Assured Shorthold Tenancy, unless the property has a rental value of more than £25,000 a year. An Assured Shorthold is for a minimum of six months; and, more often than not, these tenancy agreements run for a year. They can contain options to renew but the initial length of the tenancy will have been agreed at the outset. So will the rent payable, with it being stipulated in advance that when the agreement is for more than a year, or if there is an option to renew, the rent will be increased, probably annually and in line with any upward movement in the Retail Prices Index.

So, in short, the landlord and the tenant can be confident of the precise rental term and cost; and, with a properly drawn-up Tenancy Agreement, of the Dos and Don'ts.

These Dos and Don'ts will be included in a Tenancy Agreement produced by the experienced agent; but the first Don't is don't expect to find them catered for in an off-the shelf agreement form bought from HMSO or from law stationers. The agent's own Tenancy Agreement will be the product of experience, local market conditions and the constantly revised and up-dated advice of specialist lawyers, selected for being properly versed in residential property law.

As well as including the length and cost of the tenancy the deposit and the responsibilities of the tenant, an agent's own Agreement will certainly detail many other matters important to each individual property. The Agreement will cover pets (and the agent may well have demanded pet references), children and specific responsibilities like looking after common parts or the garden, even down to feeding the goldfish in the pond. It's all there in black and white and both parties must read it carefully before signing.

The agent will notify the mortgage lender of the new let and its terms and change the utility accounts as bills from the electricity, gas, water and telephone companies and for Council Tax become the responsibility of the tenant. Then, once the cheque for the deposit (usually the equivalent of a month to six weeks rent) together with the amount for the first rental period has been cleared, the keys are handed over and the tenant can move in.

During the term of the tenancy the letting agent will collect the rent, attend to day-to-day maintenance and repair as requested by the tenant and required under the terms of the Tenancy Agreement, alert the owner of major problems needing attention, account to him for costs incurred and pay over all balances due. Periodic but regular inspection visits are made to check that all is well with the property and that the tenant is happy—for a contented tenant is the best tenant.

Finally, at the end of the tenancy, the inventory and condition is checked against the signed report, deposit monies returned, or allocated against missing or damaged items, and the utility accounts returned into the owners name.

Then the whole process starts again, hopefully with the property only vacant for a very short time. In a well drawn-up Tenancy Agreement there is always a clause to allow the agent to start showing the property to prospective new tenants before the existing tenancy has run its course.

Rule 8: Get fixtures and fittings right and don't make false economies

Most buy-to-let tenants are young professionals or students, are single, aged 20 to 35 and with a busy lifestyle. Furnishings, fixtures and fittings should therefore be chosen for robust treatment. Spot-checks by local authorities will catch you out if you don't have properly fitted, tested and certified gas, electricity and water supplies, or if you run a house for sharers with inadequate health and safety precautions.

Lettings agents and local councils will advise you in detail. This is what ARLA says:

> At all levels and ages, the trend is towards unfurnished property to rent, sometimes known—especially outside London—as 'part-furnished' as all rental property that lets successfully comes with carpets, curtains, electrical fittings, fully fitted kitchens and attractive bathrooms. With the exception of the furniture itself, a property being shown to let for the first time should look no different from a builder's show house.
>
> The objections to unfurnished property stem from the pre-1974 era when disposable incomes were lower, the cost of furniture relatively high and unfurnished property was thought to be a charter for sitting tenants. That impression was firmly quashed by the courts 21 years ago and again by the 1988 Housing Act.
>
> The only exception to the swing to unfurnished property to rent is the very specialist central London corporate market. For the investor landlord, the change to an unfurnished market is welcome.

Properties that let quickly need no more than quality neutral decor along with plain colours for the carpets and curtains. It should all form the backdrop for an incoming tenant's own choice in furniture and the fabrics of the soft furnishings; and, kitchens and bathrooms aside, only electrical fittings should be left in place as properties cannot be re-wired at the end of each tenancy.

All of this helps the investor landlord with the Fire and Furnishings (Safety) regulations. If furniture—that is soft furnishings: the covers and fillings of mattresses, pillows and cushions—is supplied by the landlord in the course of business, all of it must comply and be properly labelled as having passed the appropriate tests.

However, whatever possessions the tenant moves in with, the landlord will always be responsible for safety involving gas installations and appliances. These must be subject to annual safety checks, with proper records kept. Regulations also cover the safety of electrical installations and appliances while common sense dictates that carbon monoxide and smoke detectors are fitted in all let property.

Letting agents have spent years mastering these complex safety regulations, the constant up-dating and changes to them and the timings for when the different sections became a legal requirement. They are not difficult to comply with, nor—for the investor landlord coming into the market—are they costly; but expert guidance is needed from the outset.

Rule 9: Always get insurance

Burst pipes or accidental fires can cause tens of thousands of pounds' worth of damage, so never skimp on buy-to-let property insurance. Also, do not be tempted to forget telling your insurer you are letting out a home—failure to be honest may lead to any claim being nullified.

Base your *building insurance* on rebuild costs, not how much the home would fetch if put on sale today. In some instances the rebuilding value will be considerably lower or higher than its market value. Older properties in the lower priced areas in the UK will cost much more to rebuild that their market value.

A typical *contents insurance* policy will give a fixed limit of, say, £3,500 to £6,000 to cover carpets, curtains, lights, white goods and some basic furniture.

A *minimum rental insurance* can be invaluable, especially in a recession where tenants may be in financial trouble and/or become unemployed. Its cost is tax deductible and it guarantees you receive the rent you are expecting, typically for a maximum of 12 months.

Rule 10: Spread your risk

If you develop a portfolio of, say, three properties or more then diversify so that you're able to spread your risk.

Do not always buy the same type of unit and, no matter what investment advisors or pushy estate agents say, do not have all of your investment properties in one scheme. If that fails, or there is a local economic problem deterring renters, then all of your income is destroyed.

Case study: Nigel Brockenhurst

Nigel Brockenhurst, 45, bought a five-bedroom detached house from developer Crest Nicholson at King's Warren, a new estate in Cambridgeshire close to a US air base.

He aims to rent it out to military personnel and their families based in this country on short assignments. He believes this kind of targeted approach to buy-to-let can produce dividends, even today.

"With all investment there is an element of risk, but if you can pinpoint a consistent market, where there will be a steady flow of tenants filtering in to an area, then it reduces that risk considerably", says Brockenhurst.

"Buying in university towns is a good example, but there is usually a huge amount of competition with other landlords.

Therefore, if you can find less well-known sources of tenant, such as Ministry of Defence sites, then that's even better.

The current market represents an opportunity for me, as there are more people who want to rent rather than buy. In the long run, the climate will improve. Ultimately, there will always be a demand for houses."

Student property—the next big thing in buy-to-let

An estimated 500,000 students begin university or higher education each autumn, the majority outside their local areas. Not everyone gets a place in halls of residence and most study in towns and cities with a shortfall of appropriate accommodation, so many require rooms or flats to rent.

That may sound a logistical nightmare for students and colleges but it can provide a silver lining for investors. The Government wants to increase the current 45% of over 18-year-olds in higher education to 50%, with student numbers predicted to rise from today's 2.3 million to three million by 2015.

On top of that, the number of overseas students has risen by 67% over the past decade, while rent levels have risen 31% since early 2005, partly because of general inflation but also reflecting the increased quality of student accommodation and rising expectations.

Lettings agents who find and vet students and then manage the property with tenants in situ typically charge 15% to 20% of the rental income but offer peace of mind. Without them you will have to find tenants, answer late night queries from potentially cantankerous or drunken students, and find emergency plumbers when a pipe bursts just as you're starting a holiday. The agent's fee is tax deductible, too.

Deal only with regulated agents in ARLA, the National Approved Lettings Scheme (NALS) or the Royal Institution of Chartered Surveyors (RICS). Some have a code of conduct and a fidelity bonding scheme which may protect deposits if the agent goes out of business.

Insist on seeing an agent's records of lettings in the location and of the type of property you own, and check the college accommodation bureau has good experiences of your chosen agent. Ring the agency, posing as a prospective tenant, and see how well it markets properties on its books.

If you rent out and manage your property yourself, you will have to advertise on local campuses, at college accommodation desks and on low cost websites.

You can obtain typical local rents from the college accommodation bureau but you must then screen potential tenants, take references and check credit records, all difficult and unreliable tasks, as many students are leaving home for the first time and have no employment record. You should also issue contracts, police them and be on hand for routine maintenance. If the students are from overseas you are also legally responsible for checking their entitlement to live and study in the UK. Malcolm Harrison of ARLA says:

> With an agent you know the paperwork is done. Professional tenants stay on average for 17 months but students move far more often, sometimes during term time if they're not happy or fall out with sharers.

"That means more checks, contracts and dealing with deposits. You don't want that if you're at the other end of the country", he says. For more information, visit *www.arla.co.uk*.

Low cost student renting websites include: *www.accommodationforstudents.com*; *www.studentaccommodation.org*; *www.homesforstudents.co.uk*; and *www.pastures-new.co.uk*.

Student House in Multiple Occupation

There is no shortage of red tape when you rent to a student, especially if you handle the job yourself instead of hiring a lettings agent.

If you let individual rooms in a house of three storeys or higher, to five or more students not in the same family, you may need a local authority licence for a House in Multiple Occupation (HMO).

Local authorities levy varying HMO licence fees (as much as £1,100 in some cases) and set legally enforceable standards on health and safety issues, fire regulations and room size (typically 6.5 square metres for a single person and 10 square metres for sharers or couples). If you need an HMO licence but do not get one, you can be fined £20,000.

Wear and tear

Common sense is the order of the day when equipping your property for several students in their late teens or early 20s, away from home for the first time. The buzz words to remember are "durable" and "low maintenance", so bear this in mind when furnishing a property.

- Key areas: Install easy-to-replace kitchen worktops, cupboard doors and vinyl floors wherever possible. Ensure shower units are not buried in walls so do not trigger re-tiling after a repair.

- Bike sheds: Students are the biggest victims of cycle theft but bikes stored in hallways are awkward and unsightly.

- Window locks: Student rooms are particularly vulnerable to theft.

- Gardens: Make it is either a plain lawn (requiring occasional mowing, at most) or do a Garden Force-style conversion to turn it into durable decking or gravel which requires no ongoing attention.

- Broadband: Students use the internet extensively so need at least standard 512 kilobytes broadband. If you fit a wireless hub they can access the web throughout the property, which will boost its appeal.

- Insurance: In case of fire or flood, let alone theft, a landlord may want to take out specific insurance through a specialist provider like Endsleigh. Premiums vary, sometimes unpredictably; it's cheaper to insure a student house in Brixton than in Leeds, for example.

Just like the rest of the market, the student sector is all about location, location, location.

If you are buying for your son or daughter, look near a city centre or a busy suburb, near a station or a good bus route. But most of all, buy in a known student area.

Student accommodation should ideally be within walking distance of the main college buildings but if this is not possible then it is essential to be near good transport links. Students prefer to live near shops where they can buy groceries on a day-to-day basis, so look for convenience stores.

Opinion is divided on whether to buy an older property (probably cheaper but requiring upgrading) or a new property (more expensive but with modern facilities and unlikely to need upgrading for a decade).

"Ignore new-builds in city centres and buy good quality second-hand homes in secondary locations. You don't pay a new home premium and don't compete in a location over-supplied with flats", says Richard Donnell of property consultancy Hometrack.

But David Pretty of the New Homes Marketing Board says new student flats often come with low maintenance layouts and even communal gyms. In the current market, developers of student flats are offering "incentives and assistance packages that the second-hand homes market could never hope to match", he says.

There are two ways to buy into the student accommodation sector. The first is to buy a typical three- or four-bedroom terrace house and convert a reception room to maximise rent-paying student tenants. Parents will often follow this route, installing their son or daughter in one room and renting the rest. The second is to buy a new apartment in a purpose-built block.

If you are buying for a son or daughter about to start college, you have no say on location once they have been successful in getting the course of their choice, but if you are investing purely to maximise returns, select a place where a burgeoning student population is not matched by a big supply of accommodation.

Estate agent Knight Frank has created an investment league table of places with 20,000-plus students (see below). It has factored in the ratio of students to halls of residence, the number of foreign students requiring year-round accommodation, predicted growth in student numbers, and the proportion of students privately renting off-campus.

Location league table:

1. Edinburgh
2. North London
3. East London
4. South-west London
5. West London
6. South-east London
7. Central London
8. Newcastle
9. Manchester
10. Birmingham
11. Glasgow
12. Cardiff
13. Oxford
14. Bristol
15. Nottingham
16. Brighton
17. Leeds
18. Plymouth
19. Canterbury
20. Southampton

How much to charge (2008–2009)

Town	Per week average
London	£102.85
Cambridge	£82.98
St Andrews	£82.29
Exeter	£77.54
Chester	£77.12
Oxford	£74.71
Brighton	£73.71
Bradford	£44.61
Wolverhampton	£43.49
Stoke-on-Trent	£42.04
Middlesbrough	£41.54
Crewe	£40.33

Source: Knight Frank, 2008.

Case study: Alan and Stephanie Piper

Alan and Stephanie Piper bought their daughter Lucy a new one bedroom apartment to help with her education—but not just what she was learning at the Anglia Ruskin University in Chelmsford, Essex.

"We wanted to help her learn about life, too. The apartment made it a lot easier when she was training to be a teacher at the university—it eased our concerns about security, but it also helped her learn how to own and keep a property and have the responsibilities that come with adulthood", says Stephanie, a director of a diamond brokerage.

She and husband Alan, a self-employed plumber, paid the deposit and mortgage on the ninth-floor flat in The Tower at 53 Park Central, a Barratt Homes scheme in the centre of Chelmsford. Lucy kept the property when she started her first job in London.

"It just ticked all the right boxes. The university was on Lucy's doorstep, there are electronic security systems, it's close to bus routes and the railway station, and because the property is modern it doesn't require much upkeep", says Stephanie.

"My apartment's amazing," says Lucy. "The first thing you notice are the French doors that open on to a balcony and the fantastic views over the town."

Stephanie and Alan considered buying an older house and asking Lucy to run it, with other students as lodgers, but felt that would have been a distraction to her studies.

Instead, they chose a block with a concierge service, secure underground parking and a residents-only gym.

Lucy, meanwhile, has left college with more than just a teaching degree. "Running her own place and now owning it means she's more grown up, more responsible, more ready for life", claims Stephanie. "The property has been part of her education, too."

Resource centre

Homecheck, which describes itself as "the ultimate guide to environmental and planning risks, neighbourhood information and what to look out for with the new Home Information Pack" is useful if you are buying an investment property (see *www.homecheck.co.uk*); the National Federation of Residential Landlords is at www.nfrl.org.uk; ARLA is at *www.arla.co.uk*; the National Landlords Association can be found at *www.landlords.org.uk*; updated rental statistics are available at *www.rentright.co.uk*.

If you manage your own tenants you may want to use credit reference agencies such as Experian (*www.experian.co.uk*), Equifax (*www.equifax.co.uk*) and Call Credit (*www.callcredit.co.uk*).

In the Appendices of this book, you will find a draft form of an Assured Shorthold Tenancy and details of the various tenancy deposit schemes that exist.

An Overseas Investment Property

Introduction

It's everybody's dream, isn't it? That gorgeous white villa in the sun or a ski chalet to escape to in the winter, or that flat with a balcony in one of your favourite foreign cities. We all want to own at least one of these, don't we?

But dreams aside, can property abroad make you any money?

The answer is a resounding "Yes" but if you think it takes a lot of research to find the right property project to make money from in Britain, it takes even more hard graft to do the same abroad.

The reasons are obvious—we possibly do not speak the language, they probably have estate agents and developers that are even less regulated and pushier than their counterparts in the UK, and objective guides like house price indices and demanding building regulations are difficult to find in many nations.

Just as you need to decide on your type of property project in this country, you need to do the same when considering buying overseas. A family villa in Istanbul is not going to attract many holidaymakers, but an apartment near one of its main centres may well rent well to students or temporarily relocated international executives; and likewise a holiday home on the Portuguese Algarve is not the purchase if you want year round long-term lets to corporate professional families.

The permutations are many and, for ease of reading and understanding, I set out below a guide to the sorts of properties available in the main countries where Britons buy and holiday. So I will talk about sunshine villas to rent in Florida in the US, for example, and ski chalets to let in the Alps.

The majority of my descriptions will be for homes that are aimed at holiday tenants seeking to rent on short-term contracts (often a week only, rarely more than a month). This is for two reasons:

- First, most people who own a home abroad want to use it themselves a little of the time at least—it saves money (so contributing indirectly to the profit making component of this investment) and it is of course good fun to have a home overseas that you can enjoy yourselves.

- Second, just as buy-to-let is saturated in many British city centres, the same applies in many major cities across the world. Literally thousands of Britons have bought buy-to-let flats in cities as diverse as Las Vegas, Paris, Talinn, Prague and Cape Town. Many of them may be making healthy returns, but the likelihood is that these places are now over-supplied with rental property—so let's not waste our time on them. It is, of course, perfectly possible to go to some secondary city in, say, Hungary and find that its buy-to-let market is roaring ahead for peculiarly local reasons; perhaps a new university has opened, or major firms have relocated to the city. But the amount of research required in finding this sort of location and the struggles of language, law and currency required to executive the purchase and subsequent rental, is simply too much for all but the most extreme investor.

Before we look at individual countries, we should talk about how to research an area overseas that may make a good investment. The technique may have a lot in common with buying in, say, a city like Manchester or a second home haven like Salcombe. But there are plenty of differences, too.

Preliminary points to consider

Let's start with *access*. Accessibility from the UK is vital, whether your investment is just across the English Channel or on a tiny island in the Caribbean.

Short haul destinations that are likely to be popular, in demand, so producing possible capital appreciation for properties, are those well served by budget airlines.

Even in 2008, when the credit crunch and the recession hit the world, the UK's 11 budget carriers launched 29 new routes. That is good news for property investors because high flight frequency, low cost, and ease of travel are key requirements, especially if an investment property is to be rented out successfully.

But remember, budget airlines can break as well as make a potential second home destination. Wizz Air and Ryanair both shut down new services at short notice when there were low passenger levels. In 2005, EUJet, which flew to 22 destinations from Kent, went into liquidation. Budget carriers Volare, Duo and Air Polonia collapsed in 2004, leaving many holiday homeowners with difficult journeys to their properties. Ryanair and Flybe have in the past axed routes, such as St Etienne and Clermont Ferrand in France, leaving new property buyers in these areas high and dry.

One way to avoid the same fate is to buy only in a location served by at least two airlines for at least two years—it's not a guarantee that there will be no problems in the future, but it may show a certain security for the destination.

Long-haul locations present different and greater problems.

For example, there are direct flights to only about half of the main Caribbean islands from London, with a few direct services from Manchester or Edinburgh, too. Barbados and Antigua are the Caribbean destinations best served for direct flights, although the shortest flights are to Antigua. But all direct services to the Caribbean are expensive: most scheduled return trips cost at least £500 a seat.

And don't forget island hopping, which can add to the journey by many hours.

If you buy on one of the smaller islands now seeing extensive development—for example, Abaco, Cancuan, Grenada or Nevis—then the journey from the UK will involve a main flight of some 10 hours (after the usual painful check-in wait) and then a further delay of possibly several hours at the destination airport as you wait for a connecting flight, often using a small aircraft of anything from six to 30 passengers which is not comfortable for some more nervous travellers.

I have been to many Caribbean islands, some taking almost a day to reach from the UK. Personally, most of them are worth the wait—they are beautiful beyond anything you see closer to home, even in the best parts of the Mediterranean. But they do take an awful long time to reach and the air fares can cost a fortune.

The next point to consider is *infrastructure*.

The Bulgarian ski resort of Bansko is a good case in point here. When large-scale residential development began in 2002 it became clear that this village, which under communism had been modernised very little since the 1950s, would require brand new roads, a sewage system, hotels, shops and, ideally, its own airport to accommodate the huge increase in property, tourists and investment. Some of that infrastructure came, some of it did not—and because the construction of new apartments was roaring ahead, with around 30% of them being sold to Britons, the local council wisely imposed a moratorium on new building for about two years while the infrastructure "caught up".

But alas then came the credit crunch and to this day Bansko is not as good as it should be—it has not caught up.

There are long queues to use the ski lifts (there is one new lift in service but at least one more is required and has long been promised, but not so far delivered), while the airport has yet to appear and roads and even electricity generation are less reliable than they should be.

No one intentionally missold Bansko—it was just that, in 2002, all of this looked possible in a short time. There was the possibility that Bulgaria would enter the EU (it did so, but later than anticipated and with many caveats requiring the end of local government corruption); an airport and new facilities were promised if it hosted the 2010 Winter Olympics (it was shortlisted but did not win the games, so those allied improvements have not taken place).

And then in 2008 the banks stopped lending, which made it hard for developers to complete schemes and harder still for overseas buyers to purchase them. A few of the early investors sold at a modest profit but many others now own properties in an area that did not fulfil its potential.

The same could be said of locations in the Caribbean. Islands that have been second home locations for longest, such as Bermuda and the Bahamas, have the best infrastructure: they tend to be wealthier, use the US dollar as de facto currency and have the largest properties. Newer destinations, such as Grenada and St Lucia, have less sophisticated facilities.

So if you are buying, look for:

- good roads, ideally already built or at least under construction
- genuinely reliable public transport
- adequate shops, from food stores to banks
- a range of entertainment facilities (and not just a disco in a club)
- hotels that have committed to an area with substantial investment
- ideally, a nearby airport that hosts international flights.

One way of guaranteeing good quality homes and service is to buy in a resort linked with a top hotel brand such as Four Seasons, Raffles, Marriott, St Regis or Ritz-Carlton. Although such flats and houses usually sell at well above average prices, they are inevitably in prime locations, have well-tended grounds with extensive leisure facilities—sometimes even private beaches, golf club membership or their own personal access to ski slopes—plus in-house restaurants, shops and transport.

The next point to consider is *local politics*. This is not as eccentric as it seems, and can make or break a location.

Cyprus, for example, saw many properties bought by foreign investors in 2006 on the grounds that an end to the long-running dispute between Greece and Turkey over northern Cyprus would lead to a reunification of the island, leading to significantly greater international infrastructure investment and wealth. That agreement did not happen and the wealth has not poured in on the scale that was expected.

As a result, there remains long-term uncertainty over homes bought in what is now formally called the Turkish Republic of Northern Cyprus (TRNC).

The TRNC authorities, recognised only by the Turkish Government, sold or gave away plots of land and houses left by the Greeks after the arrival of Turkish forces in 1974. Some were given as "compensation" to Turks who moved from the south to the north; some of these were later sold on while others were given to Turkish soldiers, and some were taken over by Turkish Cypriots or sold to foreigners by opportunistic estate agents.

As a result, many of the homes and plots that belonged to Greeks before 1974 have now changed hands. Most new owners are Turkish Cypriots but, in many thousands of cases, the new owners are Britons or other foreigners. Official records are scant, but one estate agent says 1,800 homes in northern Cyprus are lived in by Britons.

This situation imploded in the early summer of 2009 when the European Court of Justice ruled that Meletis Apostolides, a Greek Cypriot architect, should be able to reclaim land his family abandoned when the island was partitioned in 1974, and on which a British couple had since built a holiday house. The judgment could prompt hundreds of similar lawsuits. Northern Cyprus may be a lovely part of the island, and many obviously think so, but it is arguable whether the potential problems on this scale make it a straightforward investment.

Likewise in Thailand there are (usually bloodless) military coups occurring relatively frequently, and the effect on international property purchasers is quite severe. In the past decade there have been problems concerning how Britons can buy "clear title", issues over the correct ownership of properties and plots, and the ability of foreigners to repatriate their money from rental income or the proceeds of a house sale.

The problems of Thailand and Cyprus are, of course, unusual—but many foreign destinations are problematic so care should be taken to choose the best for an investment purpose, and not to choose those which simply appear the most attractive.

Then think about *value for money*.

If you are buying a modern or new property, off-plan or just constructed, compare its price per square foot with other similar homes nearby. As we have seen earlier in this book, scrupulous buyers wanting the most for their money will measure properties they may want to buy, or get the estate agents to measure them, and calculate their relative value by comparison cost per square foot.

If you are buying in a location with a large number of retired individuals or super wealthy owners, remember they will not be in a rush to sell. On Mallorca, for example, many sellers merely wish to move because they are bored with their current home; they may be marketing it at a price that is hugely more expensive than it should be, and frankly they are indifferent to whether they sell tomorrow or next year, so they are happy to wait until a gullible buyer comes along and pays up.

And remember that in some countries it is not unusual for a seller to market the same property through a number of rival estate agents at very different prices. There are no local laws against this, and some agents make a point of advertising properties with higher asking prices than other agents, simply to see, once again, if a buyer comes along without having done adequate research.

It is wise to talk to some of the big international estate agents about value for money; just call their London international offices and chat through some of these thoughts, and because these firms operate at the top end of the market around the world they will not give you the highly partisan, subjective comments you will get from agents who deal only in one or two regions and who have a vested interest in talking them up to an unrealistic degree.

Remember *social factors* like crime and poverty.

Many people feel uncomfortable even considering these issues but they are facts of life and feature far more prominently in some countries than they do in the UK, so it pays to confront these points.

One good way of identifying potential problems in an area is to see how hotels and resorts operate. On parts of Jamaica, for example, hotels offer "full service" deals only—that is not just bed and breakfast but lunch and dinner too, not as an option but as the only deal on offer. This is to maximise the time people spend in the hotel and using its facilities, minimising their risks on what are (in places) high crime areas of Jamaica.

Likewise, even two decades after the end of apartheid in South Africa, there remains tens of thousands of people living in shanty towns, often on the edge of the country's great cities. Do you want to see these communities when you go to your holiday home? Do you want your tenants to do so? Are they likely to spread rather than diminish?

In reality, South Africa's Government has a programme to get rid of these shanty towns in the next decade, but you see the point I am making—social issues can affect the perception of a location, so can therefore also affect the profitability of an investment home.

You should obviously pay great attention to the **location and type of property** you require. Start off by considering the type of location:

- A **resort property**—this might be close to, even run by the same people as a nearby golf club, hotel, beach and sailing club, or ski slopes. The likelihood is that your home will therefore be managed by the owners of the resort too.

    These days there are plenty of year-round resorts. For example, there may be a beach for the spring and summer, plus a lake and mountain walking for the winter. Or a ski resort will have the slopes for three or four months of the year and walking, boating or even sun clubs for the sunnier seasons.

- A **city centre property**—this is likely to be an apartment and, because you will be literally hundreds or thousands of miles away, it should be registered with a lettings agent to handle tenants and find cleaners and maintenance people. These days there are plenty of holidaymakers who want to rent a flat in a big city like Barcelona or Moscow, so you have this option as well as the longer-term tenancy associated with urban centres.

- A **private home off the beaten track**—this is simultaneously likely to be the best value for money, probably the most beautiful but also the hardest to look after from a distance. You have to find the property, check it out, furnish it, then either find a local lettings agent or handle the entire exercise yourself.

- A **wreck**. There is the classic French renovation exercise where the British buy a wreck and turn it into a glorious gite, but the same can be done in many other countries (and in fact, France has relatively few unmodernised homes left). Remember that wherever you do this you must source and find the wreck, the craftsmen and then the lettings agent if you rent it out, or the estate agent to sell it.

Remember, too, to consider the *availability of finance*. This is particularly crucial in the years before banks and other mortgage lenders return to their less restrictive policies.

In 2009, for example, Britons wanting to buy in Spain—which had seen the worst proportionate price falls and oversupply of holiday homes of any country in the world—found mortgages difficult to obtain.

The maximum 80% loan-to-value (LTV) that was usually available for Spanish holiday homes dropped to a maximum of 60%—and even then, that was only for "pure" lifestyle purchases. No buy-to-let mortgages for Spain were available at all according to Conti Financial Services, specialists in foreign mortgages. It was a similar problem in Dubai—70% maximum LTVs on any property was replaced with a 50% maximum on villas and houses, and no mortgages at all on Dubai apartments. The final, but by no means least important factor to consider is the local market. Because almost all of the developed world is in a recession of some kind, prices of everything from land plots to wrecks to new homes have fallen around the world.

That makes this the ideal time to buy so long as you intend to hold on to your investment for anything from five to 10 years or more, and do not have a hard deadline when you must sell—if you do, that deadline may coincide with a price downturn.

You will quite reasonably look for market information about the country (ideally the locality) where you are considering buying. Unfortunately few countries have the array of price indices that exist in Britain. You may have to resort to relying largely on estate agents—of course, they will know, very accurately, about the market but whether they tell you the exactly accurate story is something else.

The world price index produced by estate agent Knight Frank shows how many of the globe's most popular second home and investment locations were faring in early 2009. Two key points to note are that many countries had seen price falls for the year before the publication of this index, and that even those still showing positive growth over the full 12 months were seeing spectacular falls in the final few months of that year—suggesting more falls may be on the way:

	12 months to early 2009	Q4 2008
Russia	19.7%	–1.1%
Czech Rep	19.6%	0.2%
Bulgaria	12.5%	–4.1%
Slovakia	10.9%	–0.4%
Israel	10.5%	–1.2%
Colombia	7.1%	2.7%
Indonesia	5.2%	0.6%
Hungary	4.6%	0.5%
Switzerland	4.6%	1.2%
Ukraine	3.9%	–9.3%
Cyprus	2.8%	–1.6%
Philippines	2.7%	–2.2%
Taiwan	1.9%	–5.1%
China	1.6%	–1.1%
Netherlands	0.9%	–0.7%
South Africa	0.1%	–0.2%
Luxembourg	–0.3%	0.1%
Germany	–0.7%	0.0%
Lithuania	–1.0%	–3.6%
Sweden	–1.8%	–3.4%
Croatia	–3.0%	–1.2%
France	–3.1%	–6.5%
Spain	–3.2%	–2.4%
Australia	–3.3%	–0.8%
Finland	–3.3%	–3.9%
Malta	–4.4%	–1.7%
Japan	–4.5%	–3.0%
Hong Kong	–5.1%	–12.8%
New Zealand	–5.3%	–0.1%
Austria	–5.5%	–1.7%
Canada	–6.2%	–2.1%
Portugal	–6.3%	–1.0%
Singapore	–6.5%	–6.1%
Denmark	–7.0%	–5.1%
Norway	–7.5%	–6.2%
Estonia	–7.5%	–9.3%
Ireland	–9.1%	–3.3%
United States	–12.1%	–3.5%
Iceland	–14.0%	–11.3%
U-Kingdom	–14.7%	–5.1%
Latvia	–33.5%	–16.3%

Source: Knight Frank (2009)

There are plenty of specialist magazines in the UK for would-be investors to get advice from, but heed this word of warning—and it pains me, as a journalist, to have to say this. Some of the articles are badly researched, or take at face value the data and descriptions put forward by developers or estate agents who have a vested interest in trying to get you to buy a specific scheme in a specific destination.

In well-known countries with a lot of other information sources, such badly written and partial articles can be balanced by easily obtained independent data. But if we are talking about an emerging market, that one article may be the only source of information on where you may be about to invest your life savings.

So you must undertake independent research.

Go to the following websites for hard data on a location:

- *www.worldbank.org, www.imf.org* and *www.oecd.org*—these are the sites of global economic bodies but give fantastic, detailed and authoritative information on countries and regions.

- Check the better researched articles in the likes of the *Daily Telegraph* and *Sunday Telegraph* (*www.telegraph.co.uk*), *Financial Times* (*www.ft.com*) and *The Guardian* and *The Observer* (*www.guardian.co.uk*).

- The Central Intelligence Agency (CIA) Factbook (*www.cia.gov*) and the British Government (*www.fco.gov.uk*) give exhaustive information on the political stability of countries—don't be deterred just because spooks compiled this.

- Almost all national tourist bodies give serious and reliable figures on tourist patterns in their countries—where foreigners visit, how often, how much they spend and where they live in the first place.

OK. So what next?

Tourist figures show that the majority of tourists from Britain—likely to be your biggest rental market if you have a holiday home—still go to Spain, France, Italy and Portugal, in that order of popularity. Then comes the US, Caribbean, Canada, Switzerland, Thailand, Dubai and a range of countries across Eastern Europe. To look at all of these in detail would require a separate book so we shall look only at the top four, before moving on to some general tips and case studies.

Buying in Spain

The UK Government's Office of National Statistics says some 69,284 properties in Spain are owned by Britons, although these include those owned by people still officially resident in Britain and who declare their second home to the Her Majesty's Revenue & Customs (HMRC). In reality, many more are likely to exist. A separate report from business intelligence consultancy Datamonitor says more than a third of the tourist housing stock in Spain is owned by Britons.

The best bargains and cheapest homes are not always the best investments. There are literally tens of thousands of new-build homes—mainly flats but some houses—and by mid-2009 these were typically 45% below the prices being asked in mid-2007. But they probably have some way further to fall.

Even if you buy them, are they going to rent out? All new homes in the prime locations—with front line sea views or the best mountain locations, or in the white villages—sold many years ago, so any new-builds today are often in poor secondary locations. On mainland Spain, build quality is often suspect too.

Even if you grab a bargain at a low sale price, with such a glut there is little guarantee that these prices will appreciate significantly in the near future. And as for renting them out in the interim—well, there is a vast oversupply of homes to rent, many in better locations. You may strike lucky and get a lot of renters but the odds are that you will not.

Spain is therefore a good buy only if you are happy to hold on to your property, possibly without earning much rental income, for at least 10 years or possibly more.

If you do buy, these are points to note:

- Although there are many English-born estate agents operating on the Costas and now in some cities, most agents away from the densest tourist areas are unlikely to speak English.

- Their websites are not as sophisticated or accurate as in the UK. Often a property will be registered by its seller with different agents, sometimes at different prices, so check around to get the best deal.

- Once you find a home you like and agree a price, the buyer's conveyancing solicitor must establish that the property has clear ownership title and is free of debt (any unresolved debt at the time of purchase falls to the new owner).

- Both parties then sign a contract containing local government search details, purchase price and date of completion if it is a new home. A deposit of 10% to 15% is normally paid by the buyer. If the buyer pulls out he or she may lose the deposit; if the seller withdraws, he or she must reimburse twice the total of the deposit. On completion day buyer and seller sign the title deeds, or Escritura de Compraventa, and the remaining 85% or 90% of the purchase price must be paid.

- There remains concern in the Valencia region of South-eastern Spain that, despite recent law reforms, developers still have powers to redesignate land from rural to urban. This allows them to compulsorily purchase properties in the way of larger scale developments. It is vital for a bilingual solicitor to check title deeds if you buy there.

- Always allow at least 10% of the purchase price to cover transaction fees. These are likely to be broken down as: seven per cent property transfer tax, notary and Land Registry fees of about 1.5% and there may be additional local taxes to cover utility switch-on. If you buy a new-build home from a developer you may pay up to 12% in VAT and related taxes.

Buying in France

The traditional French property market offers poor pickings to investors as it is expensive to buy into, tightly regulated and highly taxed. But leaseback schemes, introduced by the French Government to encourage more "beds" for tourists in hitherto under-performing areas, offers investors good returns. In these schemes, almost always blocks of apartments, buyers are obliged to "lease back" their homes to the developers, or an approved management company, to let out on their behalf. Two lease options are available to most buyers:

Lease with rental income

- The buyer does not pay the usual 19.6% VAT on the property.
- There is an annual rental income, often quite small—typically 4% of the purchase price—for the duration of the leaseback deal (typically 11 years).

- The buyer can use the property for a short time, often off-peak.
- Buyers must pay a service charge, plus the taxe foncière for local council services.

Lease with price reduction

- The developer reduces the purchase price to the equivalent of the rental income in the leaseback period.
- The owner can use the property for a short time, often off-peak.
- Buyers must pay service charges and taxe foncière.
- Terms vary from developer to developer but usually owners can sell at any time. At the end of the 11-year term they can enter into a new lease agreement or opt out and retain full use of the property. In some resorts in the French Alps, where authorities have a shortage of rental properties, it is obligatory to negotiate an additional leaseback deal beyond the initial 11-year term.

So far, little is known about the long-term market for leaseback properties—do they hold their value over 11 years or more? Are they popular with tourists and therefore have long-term rental potential? Time will tell. If you buy a "mainstream" property without a sale-and-leaseback deal, remember to take the following points into account:

- You have to pay up to 10% for the notaire, agents and other local fees.
- There is also a land tax, a tax levied by the local authority, and a wealth tax if your property is worth over 800,000 euros.
- Capital gains tax (CGT) is payable if you sell at a profit, although you can offset against it many of your renovation and extension costs.
- From the sixth year of ownership onwards, the CGT liability is reduced by 10% annually so if you do sell in the distant future—say 10 years from purchase—you may be eligible for only half of the CGT.

Can a Paris pied-a-terre be profitable?

Apartments are marketed by the number of rooms, not bedrooms, and prices are determined by location, size and floorspace. Anything inside "le Peripherique", the inner ring road, is defined as central Paris, and attracts a premium.

The first arrondissement is the main tourist zone and closest to where the main Olympic Games facilities were to have been built had the city won the right to hold the 2012 event. But the most fashionable and oldest neighbourhoods for buyers include the third and fourth which form the Marais. The 17th-century mansions have been split into flats, and some quarters have become mostly Jewish, others gay and alternative. The fifth is the Latin Quarter, while the sixth, St Germaine, is becoming heavily gentrified and expensive.

The 11th, 12th and 13th are historic residential areas near the Bastille, the Bois de Vincennes and Chinatown, respectively. The 16th is a wealthy area dominated by the Bois de Boulogne and the 17th is also pricey and sits close to the Arc de Triomphe and more Olympic sports events.

Almost all apartments rent well and many agents specialise in corporate tenants who want one- to six-month tenancies, or well-heeled visitors who want a holiday flat for a month. Arranging your own time in your flat will not be difficult.

Buying in Italy

For years the country has been a favorite British holiday location, but until now Italy's property prices have deterred all but the wealthiest buyers. The UK Office of National Statistics says only 3,000 Britons own homes in Italy and while this is likely to be a dramatic underestimate (the figure is of those who pay tax on the rental income or when they sell), it is nonetheless well under 5% of the number of Britons said to own in Spain and barely 8% of those who own a home in France.

Homes on sale fall into three groups.

The first consists of wrecks—those classic period properties that need restoration, and this is where the hard-working investor can easily make the highest profit.

In Italy, unlike in France, there are few property-finder services or Britons who speak the language. As a result it has previously been left to individuals to try to find those perfect restoration projects, but very few have even tried. Renovation projects are big business in Italy as the Government gives tax breaks to owners using local builders.

The second category of cheap Italian homes consists of large period houses split into smaller units. Again, this sector is becoming more popular in rural areas because it offers higher profits to developers, and because small houses and flats attract a large pool of potential buyers.

The third low-cost category consists of brand new properties, usually in urban areas. These are likely to be suitable only for long-term renters. Income from buy-to-let has historically not been as high here as in France and far short of that in Britain. So don't expect to make money from letting—Italy's private rental sector is one of Europe's smallest and strict controls mean rents from long-term tenants rise by only 75% of the cost of living, so income may not cover mortgage costs.

If you do buy, bear in mind the following:

- Buyers usually make a verbal offer to the seller as a written offer is legally binding.
- Once a price is agreed, seller and buyer sign a compromesso before a lawyer—a legally binding contract stating purchase price, date of completion and mortgage details; it may include a deposit of between 5% and 10%.
- Six to eight weeks later the completion (or atto) takes place, requiring the rest of the payment plus bills for property tax of up to 0.7% of the purchase price, and up to a whopping 15% of purchase price for fees to lawyers and agents.

Buying in Portugal

The Algarve is, of course, the most popular area for second home investors and holidaymakers alike. It is more expensive than most parts of Spain, but also less heavily-developed and generally has a higher build quality. It is renowned for its golf courses, good infrastructure and now a lot of marinas and resorts too. In short, it's classy, but expensive.

If you do buy, bear in mind the following:

- The process is relatively slow and more expensive than in most parts of Europe. First, you must get a tax card and fiscal number and must nominate a Portuguese postal address (banks and estate agents offer this service).
- Second, few Portuguese finance houses lend to foreign buyers, so most Britons arrange a mortgage at home or take equity from their principal property.
- Third, many apartments are community-run; buyers must purchase a share in the organisation and be responsible for communal areas. This makes renting out more fraught.

- Legal fees can be 2% of value; deed registration is another 1%; a thorough survey costs 1,000 euros; turning on the utilities can cost another 1%; mortgage fees are often 2%; stamp duty can be as high as 7.5%; and transfer tax on second-hand homes can be as high as 10%.

If you fancy buying a property for a long-term let, Lisbon is the obvious choice in Portugal. It has only 1.3 million residents, but it is one of Europe's great capital cities. Cobbled streets, funiculars, monuments and architecture spanning the 18th to 21st centuries dominate the centre, which is becoming better known to Britons thanks to budget air links from London and provincial UK airports.

Housing is cheap. No Portuguese price index exists, but most central flats are now below 100,000 euros while houses come in at 250,000 euros, according to estate agents in the city. The cost of living is well below that of neighbouring Spain.

But Lisbon still lacks appeal because it is, even today, relatively poor. In the past 20 years it has undertaken Europe's biggest house-building programme because of the appalling state of most homes; around 45% of its properties have been built since 1980, while the population has risen by just 6%.

But many older homes in the city centre have not yet been modernised, and are very cheap. It is possible to get two-bedroom apartments in scruffy period buildings there for 50,000 euros or less, although now these will be in unfashionable suburbs.

Vast infrastructure improvements, mostly EU-funded, have taken place in the past decade. Sewage and water systems have been upgraded, as have roads and public transport linking the city with the rest of Portugal and Spain.

Many who buy in the city therefore choose slightly off-central suburbs such as Rego, near the university and north of the centre, where apartments are typically 30-years-old. Others go further out to what locals call the Lisbon Coast, from Ericeira to the north to Sines in the south, and including Estremadura and Ribatejo.

Taking a punt on Germany

At first sight, Germany's feeble housing market performance seems inexplicable. Prices fell from 2000 to 2006, while across the rest of Western Europe values had risen by an average of 90% and those in Britain by 120%.

Three problems have held Germany back, at least until now. First, supply of homes has far exceeded demand. After reunification between East and West in 1990, there was ferocious new house-building by developers, vying for buyers with hundreds of thousands of unloved Communist flats suddenly on sale.

Second, subsidised housing and strict rent controls have made renting much more attractive than buying on the open market. German owner-occupation is only 43% (it is 70% in the UK), while in Berlin it is a mere 12% and in Hamburg 20%.

Third, the cost of buying is high (you pay about 12% of the purchase price in fees) and it's not so easy to get a mortgage in Germany as in most of Europe.

As a result, prices remained low. But just before the credit crunch there were changes appearing in the market and many German economists believe these will reappear when the global recession ends—so getting into the market now may be a shrewd move.

A few pioneering British investors have already taken the plunge. London-based civil servant Simon Luker has bought a small Berlin studio apartment in a modernised period block in the suburb of Neukollen for only £21,000. "I've been looking to invest overseas for some time", he says, "and was surprised to learn German property was considerably lower than even Riga or Prague".

Ray Chapman, a business manager from Hastings, and Sue Hart, head of a health treatment centre in Hertfordshire, looked at Eastern Europe but decided against investing because the infrastructure was relatively undeveloped. Instead they bought a £35,000 one-bedroom apartment 15 minutes from Berlin city centre. "Berlin has a superb infrastructure and is almost tailor-made for buy-to-let", Ray explains:

> Our investment is long-term, as German law dictates that any property sold within 10 years attracts 25% capital gains tax. Berlin is an attractive rental market that isn't based on seasons or low-cost airlines getting routed.

An unusual angle

If you want to try something unusual, try "fractionalising" your holiday home so you can allow it to be "bought" by between three and 12 other people on the usual fractional basis of dividing its deeds and ownership into quarters (so allowing each co-owner to have three months usage per year) or 13th shares (so allowing each co-owner four weeks usage per year).

Case study: Charles North

Charles North makes his own pizzas in the restaurant he runs at Bedale in North Yorkshire—but it's dough of a different kind that he is hoping to make from a two-bedroom apartment he owns in the centre of the Czech capital, Prague.

He bought the property near Wenceslas Square in 2005 with his wife Daniela and son Chris. They are typical of a new breed of British buyer for whom income and potential capital appreciation are more important than sunshine.

"We researched extensively, originally considering other east European locations such as Warsaw and Krakov, but we ruled them out because they didn't get as many visitors and didn't have such good air links as Prague", he explains.

"Buying the property and renting it out was fairly straightforward, as there are English-speaking lawyers and letting agencies in Prague. We've used the property a little ourselves but it's primarily an investment. It's got an occupancy rate of 55-60 per cent so we're very pleased with how it has gone." The current rental more than covers their mortgage and the couple plan to sell the flat in four years' time. Then, they say, they may buy another property in another emerging location.

The website that came up with this idea, *www.nuwireinvestor.com*, advises that a property management company must be set up to ensure the varying owners are each serviced with repair and tenant management during their "fraction". The website says there are 10 rules to adhere to if you want to follow this route:

1. Evaluate the home: Consult a good interior designer in whatever region of the world your property is in, to give an estimate on bringing the home and furnishings up to "like-new" condition.

2. Evaluate the location: Consider how prime the location is for someone on holiday. How many attractions are within a short walk or drive of the property? How many seasons of the year do tourists visit the area?

3. Check the legal situation: Are you entitled to turn the property into a fractionally-owned home? Consult a lawyer.

Case study: Johnny and Louisa Paravicini

Johnny and Louisa Paravicini can teach us all a lesson when it comes to making full use of a holiday home. An occasional summer week away at a place on the coast, or a one-off winter ski trip to a traditional mountain lodge is not for them.

Instead, they make year-round use of their holiday apartment on Lake Geneva because they chose the perfect location in which to sail, walk and cycle in summer and enjoy a range of snow sports in winter.

"The apartment overlooks a marina on Lake Geneva directly opposite Montreux", says Johnny, a consultant who lives in Hampshire and advises international clients on moving to and working in Britain:

> It's a vast piece of fresh water so we sail around, then anchor up and swim. In the winter it's just a 25-minute drive to the ski resort of Villars. It's a fabulous place to have a base.

To cap it all, their children and grandchildren use the one-bedroom flat too. "So it's intensely occupied. A perfect investment", says Johnny.

The family is amongst a new breed of clued-up holiday homebuyers who don't want to see their property empty for much of the year.

"Year-round appeal is particularly important for holiday home and investor buyers", says Sean Collins of Pure International, an estate agent:

> Projects that can deliver year-round rental occupancy offer greater protection for investors—and are positive for the local economy.

Holiday Rentals, one of the largest lettings agencies for British-owned homes abroad, says the most popular locations are the Canary Islands, Paris, Rome, Barcelona and New York.

"The common denominator is they're all year round", says Holiday Rentals' Sarah Chambers. "Some 60% of our clients own one or two foreign homes. They almost always rent them out for the whole year. That will work where there are cultural or sporting attractions", she adds, suggesting owners should make it clear in advertising that there are things to do all year round.

4. Prepare your property: Have the appropriate refurbishing completed and focus on the details that make a home comfortable—good quality white and other furniture, luxury materials, high-speed Internet access, landscaping, and so on.

5. Arrive at your selling price: Consult a local estate agent, especially if the agent has experience of fractional ownership. A rough guide is to multiply the local market value of by 0.5 to 1.5 times to determine the total price for your home. It will be well in excess of the "normal" market price because as a fractional property it will get intensive use and high levels of wear and tear by many families (more even than a traditional holiday home that may well be empty for much of the year). .

6. Determine your fraction size: Set the fraction size based on your research of vacation patterns in your area and the price-per-fraction you wish to achieve.

7. Create a fractional use plan: Research use plans so that you have a clear idea of all of the variations—four co-owners in total or 13? Select the one that makes the most sense for your particular location and your target buyers.

8. Develop a management plan: Set up the homeowner's association and arrange for property management and concierge services. Calculate the yearly expenses so you can estimate each owner's yearly fees.

9. Develop a marketing plan: Develop a diverse marketing plan including direct mail, e-mail, print and online advertising. Consider what incentives you can offer, such as an exchange programme with other fractional properties in other parts of the world, owned by yourself or any of the other co-owners. Experts advise spending 3% to 5% of the total value of the home on initial marketing, then 1% to 2% each year thereafter.

10. Perfect your sales pitch: Develop a plan for handling sales calls, showing the home, closing the sale, and handling deposits and down payments.

Inspection trips

The hotel foyer is teeming with British visitors eager to banish the winter blues with three days under the Costa Blanca sun. A holiday? No. This is house-buying, 21st-century style.

Fifty seven of us are here on the first evening of an Inspection Trip, a frenzy of visits to properties on sale around Torrevieja in southern Spain, arranged by the estate agent Atlas International. These subsidised trips were once used to sell only low-cost homes, but some buyers today have £500,000 to spend.

We have each paid just £49 and in return get flights, accommodation and waistline-expanding meals in marina restaurants. We inspect up to 12 properties daily, enough for Atlas to hope that we will eventually put a 10% deposit on anything from a studio flat to a six-bedroom villa.

My compatriots, ranging from investment buyers in their late 20s to potential émigrés in their 60s, signed up at exhibitions in the UK and, according to one Atlas sales representative I overhear, they are "hot to trot".

At the height of the market in 2006, Atlas hosted two trips each week, some with a whopping 150 participants. Most are British, with a sprinkling of Germans, Dutch and Scandinavians. Trips take place all year round with only brief respite at Christmas. They can be hectic, according to the sales reps, especially if buyers have no specific property in mind.

About 100 different house types from 14 developers are available for us to view, both on the coast and in the hills, and mostly in surprisingly dense estates.

The crunch comes on day four when, after yet more visits for those not suffering from floor-plan fatigue, it is make-your-mind-up-time. To buy or not to buy?

It is quickly apparent that an inspection trip is both soft- and hard-sell. Soft, because it is like a mini-break with property visits between generous meals with wine; hard, because the pace is brisk and the reps are very, very keen.

Cheesy though it was in parts, it was easier and more fun than solitary treks to different (and sometimes indifferent) estate agents, and occasionally fellow-trippers make it even more interesting.

But this pack-them-in, keep-them-busy technique is not for everyone. If you're in the market for a crumbling Spanish finca with its own orange grove, inspection tours are not for you—they look solely at new-build homes, where the biggest profits lie.

Equally, some people find the format too restrictive. Midway through my trip one couple hired a car to hunt down homes in other areas. They were dropped by Atlas and had to find their own accommodation for the rest of their stay.

I judge that at least half have signed up. Atlas expects that the figure will be nearer three-quarters, but I am not convinced. As for me, in true journalistic style, I made my excuses and left without buying. Then I flew home for a rest.

Resource centre

It may not always seem obvious, but Britain is advanced when it comes to technology—at least where property websites are concerned. Even sophisticated nations with many international sales such as Spain, France, Bulgaria and Dubai, have few agents' websites and those that exist are often out of date.

So *www.enormo.com* is invaluable—it is Europe's largest property portal, has good navigation and great maps of homes for sale. Try bizarre searches such as "one-bedroom flats in Croatia" and it usually delivers. Likewise, *www.globalpropertyguide.com* has excellent news on national markets and prices, plus investor information on yields, taxes and laws in different countries.

Guides on how to buy in different countries are on *www.buyassociation.co.uk* and *www.worldproperties.com*. If you want to just give up and pay for someone to find a home for you, *www.relocationagents.com* lists good overseas buying agents.

Building a new life overseas once you have a home is a lot easier thanks to *www.expatica.com* (for Britons moving to mainland Europe) and *www.direct.gov.uk/en/BritonsLivingAbroad* (for tips on taxes, pensions and how to gain residency overseas).

Spain remains Britain's favourite holiday home location. Impartial information is hard to come by but *www.kyero.com*, a sales site, has good price data and *www.spanishpropertyinsight.com* gives neutral views on local markets.

Wherever you buy, you need independent advice; *www.lawoverseas.com* has good legal information as does the Law Society at *www.solicitors-online.co.uk*, and Goldsmith Williams at *www.goldsmithwilliams.co.uk*. For insurance help try Saga at *www.saga.co.uk* and Andrew Copeland at *www.andrewcopeland.co.uk*.

For mortgage help try Conti Financial Services at *www.mortgagesoverseas.com* and *Propertyfinance4less* at *www.property finance4less.com*.

Finally, if you want to buy an overseas property with cash from the UK, you will need to convert your money into the local currency. This can be expensive if you use a high street bank, so try currency specialists who usually offer a better rate, do not charge fees and provide the option of taking out a "forward contract" so you can fix your exchange rate and not have to worry about currency fluctuations. Try these: Caxton FX at *www.caxtonfx.com*; Currencies Direct at *www.currenciesdirect.com*; Foreign Currencies Direct at *www.currencies.co.uk*; HIFX at *www.hifx.co.uk*; Moneycorp at *www.moneycorp.com*; Travelex at *www.travelex.com*; or Capital IFX at *www.capitalifx.com*.

Commercial Property

Introduction

For the British general public, force-fed in recent years on a diet of TV shows, books and articles about redeveloping and redecorating homes and the long-term option of using a home as a pension, the idea of commercial property may seem incongruous.

Some people may not even know what the term "commercial property" really means, and that is hardly surprising. For reasons we will discuss in a moment, the vast majority of commercial property investment is done by institutions (banks and other financial bodies, property funds, and sometimes even government-backed investment vehicles known as Sovereign Wealth Funds).

But there is scope for individual involvement, so let us clear up simple definitions. Commercial property consists of the following:

- **Retail units**—these range from individual shops through to large department stores and even large shopping centres as well as out-of-town discount "villages". Just over one quarter of all British retail units are in London and the south east of England, and just over 85% of all units are in urban areas.

- **Offices**—these range from a small space above an old shop, a more functional unit in anything as varied as an office block or an out-of-town conversion, perhaps in a rural setting. For large scale investors, of course, it can also be an entire block or even an office "quarter" in a development. Over 50% of British offices are located in Greater London.

- **Industrial units**—these units are generally cheaper per square foot than either retail or office units, because they are often little more than vast sheds. Some industrial units are highly specific (for example, mineral quarries or vehicle manufacturing) but most are actually simply storage warehouses.

There is also the farmland category, which consists largely of land rather than buildings so is not usually categorised as "commercial property". For ease of understanding, the rest of this chapter will have, within the term "commercial property", only retail, office and industrial—but we will return to farmland at the end.

So we can see that commercial property is rather less "personal" than residential property. It is less easily understood; it possesses a language and processes that are simply less familiar to people than those of houses and flats; perhaps most critically, it tends to be much more expensive and its market fortunes more volatile even than those of the residential sector.

Yet it is worth a look. Why?

- **Returns have been good** over the long-term. If you combine capital appreciation and rental yield for commercial property, and look back at its progress since the 1970s, the typical return has been about 9% per year. This makes it at least as good as equities and gilts and possibly as good as residential property.

- **Commercial leases are long**, usually existing for 10 years (sometimes even 20) with a five-year break clause allowing tenants or landlords to part company if they wish. Compare that with, say, a troublesome residential buy- to-let where you may get new tenants every few months if you are unlucky. The thought of a sensible tenant existing in your property for five years or perhaps much more is very comforting.

- **Rents are usually paid quarterly and in advance**, which is better for the landlord's cash flow than the "in arrears" payments common in the residential sector. There is also a culture in Britain of upward-only rent reviews for commercial properties. Now on the one hand there has been a false belief amongst many that in the commercial sector, periodic rent reviews can ONLY result in rising rents—this is not the case. However, on the other hand rents have risen until recently, although the current recession is seeing negotiations between tenants and landlords producing sharp reductions.

- There is **usually no need to pay for a management company or lettings agent** once a tenant has been located and is in place. Although there are exceptions to the rule, most commercial tenants are more reliable than residential ones, according to experienced investors. Major companies and franchise operations obviously have codes of conduct that each individual unit must adhere to; even small single-unit companies tend to have good relationships with landlords and are reliable. It is also expected that the tenant undertakes and funds maintenance during the course of the tenancy.

- It is widely understood that contract law that relates to commercial property is **skewed in favour of the landlord**—again, the reverse of the residential sector, where the law is structured to avoid tenants being forced into homelessness.

Put like this, commercial property seems a great sector, and it is perhaps a mystery why more small-scale investors are not involved. But there are some challenges—these are not disadvantages as such, but they can make the going tough for "amateurs" without large-scale financing.

First, *research can be very difficult.*

There are few equivalents of, for example, the Nationwide or Land Registry house price indices to guide you on prices, so "armchair research" can be hard. Although the Investment Property Databank is a useful organisation and publishes frequent market reports, these tend to be about "top level" commercial property which attracts large scale investors—there is a lack of data for the newcomer to access. There are also few High Street agents that sell commercial property (although there are a few), so you cannot even wander into a local agency and chat through options.

Second, *the market is quite small and tough to enter.*

This small number of agents is one manifestation of this, but the size issue affects the whole market—commercial property in Britain is worth about £500 billion in total, only one eighth of the size of the residential sector. That means there are fewer properties around to begin with, especially when you consider they change hands less often than houses and flats.

Third, *you need a hefty sum to begin investing.*

Mortgages, even before the credit crunch, were harder to get for commercial properties and things are obviously tougher in these recessionary times. Few lenders would now give anything more than 65% loan-to-value (that is, a loan equivalent to 65% of the property's asking price), so you need to be able to get a 35% deposit easily. Also, commercial mortgages have interest rates that are typically 2% or 2.5% above the Bank of England's base rate, so are not as cheap as residential loans. For these reasons, some experts say you should have a clear £100,000 to invest before considering going into commercial property.

Fourth, the commercial property market is characterised by *low liquidity.*

This means that, even when the market is good, commercial properties often take a long time to sell. One problem is the length of the leases—the very reason that makes them good for a landlord who is running the property, makes it tough when the landlord wants to sell up.

Fifth, *the market can be volatile.*

We all know that house prices fell 20% to 30% in 2007–2009 but that was highly unusual. In the commercial property market, values fall (and rise) that much in a year, and it's not unusual for that to happen three or four times in each decade.

Sixth, commercial property has *complicated planning arrangements.*

In truth, these are probably no more complex than their residential counterparts but, of course, most people are not familiar with the commercial equivalents—so they appear forbidding. In the Appendix to this book there is a list of different commercial property "use classes" defined by planners (you do not need to have intimate awareness of this, but it is useful to see the options and to become familiar with "classes").

So if you are not deterred by these challenges, try the following approach if you want to get involved in commercial property:

- Location is key. Small, cheap retail and office units are very heavily occupied, especially right on the edge of town and city centres. Visit some local schemes—old or new—and see how well they are used on weekdays, weekends and evenings if appropriate.

- Register as a buyer with an agent. In the commercial property world, selling agents also work closely with buyers, and they effectively try to "match" unit and purchaser. Auction houses also sell commercial lots, and this method of buying has become much more popular in recent years—partly because auctions also have finance providers in the room, so "one stop shop" deals can be made.

- Bear in mind the transaction fees associated with a commercial property, which can be expensive primarily because of the difficulty of due diligence. To buy, the fee is usually 1.7625% of the purchase cost, plus VAT, for the agents' advice during the transaction, which may include valuation, building surveys, environmental audits and mechanical and engineering tests. Other fees include Stamp Duty Land Tax which is currently 4% for units over £500,000, 3% for prices from £250,000 to £500,000, and 1% between £120,000 and £250,000.

- Avoid fringe locations in tertiary towns where 15% of shops are closed, especially in a recession. If more than one or two more shops in these secondary locations close, then the entire fringe may be at risk. Retailers move in packs and these locations will be the last to recover from a downturn.

- The biggest gains—and risks—are to be found in new developments that have no occupiers yet. Buying an empty shell and getting an agent to find an occupier is undoubtedly the best way to make money commercially.

- Restaurants and bars often fail (not always because of ineptitude on the part of the owners or managers), so beware of buying units with these occupiers—they may simply go bust and leave you with a vacant unit.

- When you buy a unit with a tenant in situ, take a realistic view over rents. In the current financial climate, the rent you receive from a sitting tenant may have been negotiated some years ago when the economy was stronger—the next time rents are renegotiated, they may have to be lower to keep the tenant in business and to keep your property occupied.

There is one more area of commercial property to consider—the *indirect* aspect.

By this, we mean it is extremely straightforward to buy into a form of fund or investment vehicle that gives you a stake in the commercial property market without owning a specific unit. Before considering this in detail, you would be advised to consult an Independent Financial Advisor, but your choices include Life and Pension funds—there are about 20 of them—or Unit Trusts or Real Estate Investment Trusts (REITs).

REITs are less well-known than most forms of indirect investment; they are bodies with a special tax status and do not pay corporation tax on income or capital gains from their property investments, making them highly tax efficient. Most of the UK's larger quoted property companies have said they will convert to REITs at some point in the future, although the combination of the recession and some uncertainty about future legal changes to REIT status has delayed this trend.

When you sell your funds you pay a hefty fee—usually just over 10%—so getting professional advice is absolutely essential.

A digression, but an interesting one: farmland

British farmland has long held an iconic and affectionate status but it is now a serious asset class and one that matches many rival investments and even outstrips them in recessionary times.

For example, the average price of British farmland rose by 21% in 2008 to £4,200 per acre, despite signs of a slowdown late in the year. This is a remarkable performance compared with, say, mainstream residential prices which saw falls of some 15% in the same period. It takes the five-year increase for farmland to 135%, well ahead of houses and also the "conventional" commercial property market we have been discussing so far in this chapter.

As a result of this strong performance, farmland ownership is expected to spread beyond its traditional boundaries; in 2008, for example, investment was cited as the primary reason for buying in 29% of UK farmland transactions, up from 16% in 2007, and at the time of writing, 2009 appears to be producing more of the same.

Returns from agricultural-related uses of land, including farmland and forestry, have outpaced those from other property markets, as well as gilts and equities in recent years. Analysis by Savills shows that yields for those renting out land to tenant farmers, for example, have been over 20% across the past three years.

So it is not surprising, therefore, that since the start of the credit crunch most banks and building societies have taken a less restrictive approach to borrowers seeking to invest in farms than those in other property sectors, because of its track record.

Relatively straightforward access to funds means that in 2008/ 2009, farmers accounted only for only 53% of purchases, meaning 47% came from other sources. Some 23% were "lifestyle" buyers who wanted a country home to live in, while 13% went to existing private land owners who were after additional fields or paddocks; both of these groups are likely to have seen the land as an investment.

The remaining 11% went to corporate and institutional investors, who chose farmland above other assets.

With lifestyle buyers likely to be thin on the ground in the future—many of these have historically been recipients of handsome city bonuses, which will not be seen in their usual number this year— there is likely to be more opportunity for new investors.

Certainly a lot of foreign buyers think so, as Irish and Danish purchasers in particular have soared in numbers recently. Partly it is, perhaps surprisingly, because land prices are not as high in Britain as elsewhere. Another reason is that Britain does not have the scale of bureaucracy seen elsewhere—believe it or not—and also because the exchange rate movements of the past year have created real agricultural bargains for foreign buyers.

Farmers in Denmark and Ireland, both north and south of the border, face high local land prices. For the Danes there are also extensive restrictions over the volume and location of additional land they can buy nearby. Therefore, farmers from both of those countries look enviably at Britain's relatively farm-friendly climate and regulatory regime, and have arrived with us in increasing numbers.

For new investors, farmland has some clear advantages.

There will always be demand for land from those who want to buy good farming locations, or who want to increase privacy around their existing farmhouse home by purchasing additional land. If government housing targets are confirmed when the current recession ends, so it is expected that some agricultural land may be used for new homes too—giving a sharp increase in value, should it happen.

There are strong alternative investment prospects, too, thanks to the likely expansion of the use of bio-fuel currently being trialled by airlines and some car manufacturers.

Industry experts predict that this will help create a constant and reliable demand for crops, either for consumers to eat or as components for bio-fuel. This gives a level of certainty for investors that has occasionally been lacking in farm production.

Like every other asset, farmland faces challenges and worries in times of recession. But history suggests it is well placed to cushion current and future investors from the deepest troughs or most unsustainable peaks that have characterised other sectors.

The decreasing numbers of farms that have come on sale in Britain in recent years have also contributed to its scarcity value. As the old cliché goes, when it comes to land, they aren't making any more of it.

Britain's regional farm prices

Agricultural land has appreciated well in recent years, in contrast with, for example, residential land and homes, which in some cases are returning to their 2004 levels.

Region	Price/acre	Five-year increase to 2009
Scotland	£2,928	+112%
North of England	£3,932	+129%
East Midlands	£4,180	+110%
Wales	£3,557	+182%
West Midland	£5,173	+152%
East of England	£4,796	+123%
South East England	£4,978	+158%
South West England	£4,401	+129%

Source: Savills Research, 2009

Advantage agriculture in recessionary times

Agri-related assets have been extremely profitable in recent years and have held their own with many other asset classes over the long-term, too.

Asset	3 years	10 years	20 years	30 years
Let land	23.4%	17.0%	13.9%	9.9%
Farming	20.1%	7.1%	9.7%	11.1%
Forestry	24.0%	7.0%	n/a	n/a
Let residential	5.4%	3.4%	12.0%	15.7%
Commercial	−3.0%	7.8%	7.5%	10.0%
Equities	−6.9%	0.5%	7.9%	12.2%
Gilt	4.7%	5.0%	8.6%	10.4%

Source: Savills Research, 2009

Case study: Michael Millington

Michael Millington turned to commercial property when he saw the first signs of the residential market downturn in late 2007.

"I had three buy to let properties already—all houses, let out to forces personnel on the outskirts of Oxford—but recognised that there wasn't any point expanding that portfolio at the top of the market in 2007" says Millington, who is an accountant who, as he puts it, "enjoys dabbling in bricks and mortar".

"So I spent about six months researching local shopping centres in secondary towns like Kidlington and Towcester. In the end I went for a fruit and veg shop with a sitting tenant. It cost only £88,000 at auction, and I've since spent about £1,100 on essential work. My rent is about £600 a month so the yield is good" he says.

Millington admits he expected the commercial market—which by 2007 had already been in the doldrums a year—to have picked up more rapidly, although he is not planning to sell for at least another five years:

> It's a good steady income with a reliable, always-in-demand tenant who hasn't suffered much in the wider recession. I suspect it's never going to be a spectacular investment in terms of appreciation, but it's the most stable part of my portfolio right now.

Resource centre

The websites *www.reita.org* and *www.ipf.co.uk* offer interesting guides to the commercial property world, while the best data source is on *www.ipd.com*. For regional information go to *www.propnews.co.uk* and *www.commercialpropertynews.co.uk* which gives a daily digest of news, but the best source of data for this sector is the subscription service *www.egi.co.uk*.

In the Appendices of this book you will find details of commercial property classes as set out by the Town and Country Planning (Use Classes) Order 1987 (SI 1987/764).

Ten Other Ways to Make Money from Property

Introduction

So far we have concentrated almost entirely on traditional, long-term "physical" ways of making property make money for you.

But there are other routes, a kind of armchair investment if you like which you can do very often from within your own home. In most cases they tend to involve smaller initial outlays and smaller gains, but most involve smaller risks too.

1. Take in a lodger

Under the UK Government's Rent a Room initiative, you can let out a room or part of your own home—it can be the entire floor of a house, for example, but cannot be a self-contained flat. The room or floor must be furnished.

You can then receive up to £4,250 a year without having to pay tax. If the rental income exceeds that you can choose between paying tax on the profit you make or paying tax on the amount of rent above £4,250.

Surprisingly, you do not need to own the property in which you live. But if you rent and wish to sublet a room, you need the landlord's consent—and realistically, landlords knowing this may rent out the spare accommodation themselves anyway.

Although students and the less well-off have long rented single rooms in family homes, tenants and landlords are now much more varied.

The fastest growing room-renting sector is among weekly commuters seconded to firms in a city for anything from three months to two years. Rather than incur the cost and impersonal nature of a hotel, they plump for someone's spare room. Typical clients work in the NHS, broadcasting, financial services and accountancy fields or on special projects for firms. The advantages to landlords, especially with "commuter lodgers", are clear and frankly they are easy money. Landlords get their home to themselves at weekends plus a substantial additional income, while the renters do not spend all week in a soulless hotel.

Tax and regulations are covered by Her Majesty's Revenue & Customs (HMRC) at *www.hmrc.gov.uk* and *www.direct.gov.uk*. HMRC leaflet RC150 entitled, "Taxation of Rents—A Guide to Property Income", is helpful.

2. Public sector leasing schemes

Public sector leasing schemes (PSLS) are great if you have a property in an area of acute housing need—many areas of London and the south east of England will qualify.

You lease your property to a housing association or local council, which then lets it on your behalf for three, four or five years. You get a fixed guaranteed rent (even if the property is empty for some of the time) and the council or housing association manages all aspects of the rental and maintenance. Most also write into their contract to ensure that they return the property to you in the original condition, so you have maintenance looked after.

The hitches? The rent tends to be about 15% below market value—although you pay no letting-agency fees, so you may actually be in pocket—and the tenants may be claimants or large families, so wear and tear might be higher than average although you will probably never know it. You have no control over the selection of tenants, and you cannot sell during your agreed lease period.

Feed "public sector leasing schemes" into an internet search engine such as Google to see what projects are operating in your area.

3. Housing the homeless

This is a variation on the PSLS theme. It is, perhaps, not a way of making your property sweat for you that everyone likes, but no one can doubt its effectiveness.

A pioneering scheme in London offers landlords a chance to rent their flats for a guaranteed income, pay low running costs and have almost no loss-making periods without tenants—as well as helping to ease the capital's homelessness crisis.

Broadway is a not-for-profit body providing support to 2,800 people per year who are homeless or at risk of ending up on the streets. It now has what is regarded as Britain's most successful scheme to persuade private landlords to rent to the homeless.

It works like this. Private landlords register directly with Broadway or with a participating local authority (currently Hammersmith and Fulham, Lewisham, Southwark and Westminster are participating, although eligible properties do not have to be within these boundaries). The charity inspects the property and vets potential tenants; those selected move in and pay rent out of their Local Housing Allowance, on contracts which can run for up to five years. The landlord gets a steady income and the tenant gets a home of their own.

Privately owned studio and one-bedroom apartments are eligible. As local authorities give priority to housing homeless families or couples to meet government targets, so Broadway is targeting single people.

It can be difficult to persuade private landlords to accept homeless clients as tenants: put bluntly, many landlords fear that such tenants will wreck their homes. "We don't do this to fail. A key objective is to give security to our clients by giving them their own front door and the guarantee of a permanent address for three or five years. But we're sensitive to the landlords and those who may live nearby. So we try to select the right people for the right property and support them as they return to living in a property over a long term" says a Broadway spokesman.

The advantages to the landlord are based on security, too. There is a guaranteed rent paid by the tenant from the tenant's Local Housing Allowance, increased roughly in line with the Retail Price Index each year, so producing a gently growing rent.

Broadway covers all maintenance and at the end of the contract returns the property to the landlord in its original condition. There will be minimal void periods without a tenant as, regrettably, there is a growing pool of homeless people registered with Broadway and local authorities.

But with lettings agencies reporting a shortage of rental properties in the open market compared to potential tenants, why should landlords go down this route?

"Simply because it's no hassle at all and a great investment" says Gary Bees, who has placed his one-bedroom apartment in Woolwich into the scheme:

> You hand over the property to Broadway, which charges an 8% agency fee—far lower than most lettings agents. You receive a rental income every four weeks so you have 13 payments a year, effectively coming from the government so there is no real worry about not getting payment. Then you get the property back as you had it.
>
> I've got 27 investment properties from Torquay to Manchester and I'm on the lookout for more. I could rent my properties out very easily through any route, but on my flat in Woolwich I clear about £100 a month profit in excess of the mortgage, which makes it very competitive with my other investments. Broadway's a very professional operation so I'm delighted with how this scheme helps me as an investor.

Broadway says it has won support from local authorities and the Department for Communities and Local Government, which administers housing policies, and so far the scheme has placed more than 80 homeless people in privately rented flats.

Visit *broadwaylondon.org* or call 0207 089 9500 for more information.

4. Extend the lease on a privately owned short-lease apartment

London's short-lease system works like this: most leasehold properties, such as flats, have leases for at least 99 years, but some of those in period buildings, built on the estates of the aristocrats who used to own large parts of central London, have run down their leases to 20 years or less. When these London estates (Portman, Grosvenor and Cadogan, for example) sell short-lease properties, they have always gone on sale at submarket prices, sometimes for as little as 50% of their market value.

Someone who bought such a flat and then lived in it for six months or more was entitled to negotiate a lease extension from the freeholder, or could buy the freehold itself. The cost of either would depend on the freeholder's willingness to sell, how much the longer lease would add to the flat's sale price (called "marriage value"), plus compensation worked out by complicated formulae involving actuarial tables and a "discount rate" based on the prevailing interest rate.

The combined costs of purchase and lease extension were usually well under the market price of a similar flat with a 99- or 125-year lease, so the buyer could make a profit if the buyer then re-sold.

There are occasional complications. For example, some buyers find it difficult to get a mortgage on a flat with a lease below 50 years because of a risk that its value could fall further if it proved impossible to negotiate an extension. Also, lease extension negotiations can occasionally drag on for years, leading to legal costs if the buyer takes the freeholder to leasehold valuation tribunal.

5. Turn your home into a holiday lettings, or buy a cottage for that market

Aside from giving you the chance to enjoy romantic weekends away, this can also be a modestly profitable business if you make the right choices before you buy.

It used to be that people would buy a place and then ask if it was appropriate to be let out, but increasingly people want to make the right investment, so check with agents before selecting an area and a property. So how do you turn a dream into a potentially lucrative reality?

- **Choose the right location**—the West Country, East Anglia, the Cotswolds or the Welsh valleys may be obvious choices, but the summer lettings season in each is only 16 to 20 weeks, plus odd weeks at other times. If you need longer letting periods to cover your costs, consider the Lake or Peak District, where winter walking takes the letting season up to 40 weeks. Some cottages can be let to holiday tenants for most of the year and to a long-term tenant in winter.

- **Choose the right sort of property**—chocolate-box cottages look fantastic on agency websites, so they let quickly, although they do not always attract higher rents than similar sized homes without traditional features. Location and outlook are key. Properties that let quickest are those with sea views, easy parking and facilities nearby. People who rent want an easy life and don't want long walks to pubs or shops. Houses sleeping four to eight people are in highest demand, and flats are the slowest to let because renters worry about neighbours downstairs or upstairs.

- **Check accommodation ratios**—if you rent to families you need plenty of bathrooms and downstairs facilities such as shower, drying room and utility space. Sufficient parking is paramount as holiday destinations become gridlocked in peak weeks and tenants need to know that they don't have to take their chances with day trippers. Outside space doesn't have to be extensive but space for barbecuing or eating outside is a minimum.

- **Fit the right equipment**—furnishings and decorations must be neutral, tasteful and comfortable. A telephone, CD and DVD players, a washing machine and dishwasher are common features. Do not skimp on comfort or you will never get repeat bookings, a traditional element of the holiday lettings industry.

- **Self-manage or hire an agent?**—as we have seen in previous chapters, many owners want to manage their own properties, but this means being on hand in the event of emergencies and always being available to handle Saturday changeovers in the peak season. If you self-manage you need to be rigorous with vetting renters—say no if you feel uneasy about letting to anyone and always ensure you have cleared funds for your bookings before giving keys or refunds. Ask for a deposit to cover any accidental damage—it's not unusual to request £100 or 25% of the rental value, whichever is the greater. If you hire a lettings agent, expect to pay around 20% of rental income as commission, plus other lump sums (typically £350 for a holiday season or £650 for a calendar year) for maintaining the property when it is empty.

- **Remember to budget for incidentals**—there is cleaning, laundering and replacement of white goods, which have a much shorter life than in a normal home, while you may feel obliged to redecorate throughout annually.

- **Get insurance**—a lettings agent will insist on seeing your insurance certificates, but even if you let the property yourself there is no point in cutting corners on this. In addition to normal property and contents cover, you must take out public liability insurance up to at least £250,000 in case little Darren gets his head stuck in your loo after slipping on a loose tile. The Association of British Insurers says litigation in this field is very, very rare; however, it is not worth taking a risk.

- **Tax benefits**—remember that to maximise what you can offset, your holiday home must be located in the UK, not be your principal residence and be fully furnished. You must make it available as a holiday let for at least 140 days a year and let for at least 70 to paying clients. It must never be let for more than 31 days to one client. If you do all that and read through HMRC's brochure "Taxation of Rents: A Guide To Property Income", you discover you can claim back some set-up expenses, including the cost of buying, altering, building, installing or improving fixed assets in the property, such as white furniture; basic wear and tear costs; some replacement items; travelling expenses to visit the property; professional fees such as lettings agents' commission, legal costs and insurance; mortgage interest—taking out a mortgage is a good move on a holiday home even if you can afford to buy one outright.

6. Fractional ownership

This is a way in which you can purchase a "share" of a property, normally a type of holiday home, either in the UK or abroad. We have already looked at how you can "fractionalise" your own existing property overseas, and selling it on in quarters or 13ths. This section refers to buying a fraction yourself.

Under the fractional ownership (FO) system, well established in America, buyers purchase a stake in a property allowing use for a set number of weeks per year. Unlike timeshares (FO's much, and often rightly, maligned cousin), buyers also receive a share of the leasehold or freehold, allowing them to benefit from capital gains.

Most schemes allow you to buy four weeks in total each year, divided either into four individual weeks (one per season) or two two-week blocks (one off-season, one high-season). The properties concerned are normally managed, often by the owners of the resort in which the property is located. In theory, you should be able to let out your "weeks" as you wish, so it could be a way of making money.

But there are drawbacks.

You cannot get a mortgage easily on a fractional property because lenders are not confident that there is a resale market, so a buyer will probably have to pull down equity from their main house. They could use that sum instead as a deposit on a buy-to-let or on a holiday cottage which, to many, would be a much safer bet.

FO is popular for properties in, say, the Mediterranean or the Caribbean where weather is relatively good all year round, but it is a different matter in the UK.

"Fractional ownership is only for warm locations" says Andrea Lee, sales manager of Poole-based Select Resorts, which sells homes overseas, including on FO basis. Christian Jensen-Broby, a spokesman for Marriott, the American hotel chain behind fractional schemes in Florida and other sun-kissed locations around the world, agrees:

> We wouldn't run fractional ownership in the UK outside of London. There's a reason we haven't opened any resort-style fractionals in this country. It's called the rain.

7. Buy a hotel room

Although it's one of the strangest forms of property investment, it could just catch on. The idea is simple: you buy a hotel room on a 999-year leasehold and someone else lets it for you.

Guest Invest—a company which has since gone into administration, it should be noted—was the first company in the UK to promote this scheme in several central London hotels. Owner Hotel has sold rooms in two four-star hotels in Hull and York, while Four Pillars has run the Cotswold Water Park scheme, where it is selling buy-to-let hotel rooms aimed at the tourist market. Developer Galliard Homes has built a 900-room buy-to-let hotel opposite the House of Commons in central London.

It sounds like a good alternative to investing in a flat or house in city centres, where there may be a saturated market for renting.

But there are risks. First, there is no established resale market to prove if hotel rooms appreciate over time. These ventures started before the recession—since when real estate prices and travelling volumes have fallen.

Second, it may be hard for investors to get a mortgage. There is little appetite amongst lenders for high-risk schemes with little in terms of a track record. When this venture started in 2004 there were a handful of lenders interested—now there are even fewer, which is why the developers tend to push their own "branded" mortgages.

A third danger is a possible slump in hotel usage. If there is a global shock—another September 11, for example—leading to reduced business and tourism, then occupancy could drop dramatically.

8. Buy into a property fund

This is the form of investment for which you would be well advised to consult an independent financial advisor—it is more of a business exercise than a property one.

Back in 2001 there were just three unit trusts and one investment trust that majored in bricks and mortar. It is anticipated that by 2010 there may be several dozen.

The reason is that Real Estate Investment Trusts (REITs) are coming to the UK, which will permit investors to shelter property portfolios in funds largely exempt from capital gains tax and corporation tax for the first time.

The eligible properties will include mixed-use schemes—for example, shopping centres with flats as an integral part—as well as pure commercial schemes ranging from business parks to out-of-town shopping malls.

Until now, investors wanting to buy into mixed-use or commercial schemes have had to pay huge sums to buy individual properties outright. REITs, which have existed for many years in countries from the US to Bulgaria, allow investors to put in much smaller sums—the rules are expected to allow investments of £5,000 and upwards.

These small sums are handled by independent financial advisors who invest the pooled contributions into a number of different developments in different property sectors.

The advantages are that your investment is relatively low-risk—it is not tied up in just one building in just one sector in just one location—and you do not have to work hard at being a landlord, finding and managing tenants and finances.

Forms of "early" REITs have been tried in the UK in recent years but they are expected to become more popular, and more attractive in their terms and conditions, after the recession. One likely reason for this is that they will become part of a new phenomenon called build-to-let.

This turns the buy-to-let model on its head. Currently, buy-to-let involves homes being built and then sold to anonymous individual investors, who let out the units. Instead, build-to-let units will only be constructed if institutional investors (and not individuals) put money in upfront. The units will be designed for long-term rental returns and not short-term capital appreciation.

The British Property Federation (BPF), which is one of the champions of build-to-let, says this new sector would attract investors hitherto buying into retail units and offices. "There would have to be some prospect of accessing capital gains from selling the property, but it would be valued predominantly for its income stream" says the BPF. Another passionate advocate of build-to-let is Nick Jopling, Head of Residential at CB Richard Ellis. He says:

> There's fantastic demand. The UK has a £400 billion rental sector, which is larger than the commercial sector. Lenders who insist on 25% deposits are pushing first time buyers to renting. And now proposed restriction on mortgages to no more than three times a borrower's earnings will inevitably provide a further boost.

The key to the viability of build-to-let is the collapse of land values in the downturn. When land prices were at their 2007 peaks investors looked to the residential sector for short-term capital appreciation because rents alone would not provide a competitive return on their hefty initial outlay. But that has changed.

"Land values are at least 50% lower than peak, construction costs are falling and local authorities more willing to negotiate on affordable housing contributions. This means Build To Let is now very viable", says Jacqui Daly of Savills. She undertook the research that formed the basis for last year's report by the Mayor of London and the BPF on how to increase institutional investment in the residential market.

That study, published in early 2008 before the recession, described the "insatiable demand" for rented property from Londoners earning £20,000 to £50,000 a year. It said owner-occupation in the capital peaked in 2001 and was now in gentle decline—a demographic pattern suggesting long-term demand for a private rent sector.

The report suggested a move towards institutional investment to forward-fund new homes and to improve lettings standards as investors sought to safeguard returns. The BPF and other build-to-let advocates point to the success of this model in the US, parts of northern Europe and Scandinavia.

So far, the only UK niche to follow the build-to-let model is a small but growing part of the burgeoning student property market.

Knight Frank, which has undertaken substantial research in this field, says between 2005 and late 2007 the total bed spaces in "privately operated student halls" rose from 7% to 10% with more growth expected when 2008 data is compiled.

This sector has seen the emergence of substantial institutional investment based on long-term rental income and not sale.

"Build to let is the only viable model for students because they need highly specific accommodation with intensive management. It's an asset class in its own right and the nearest one I can compare it with is hotels rather than any residential category" says Johnny Manns, Development Director of Hive, which has been undertaking student build-to-let schemes in London since 2006.

"Build To Let in the wider community would follow the same philosophy as the student sector" says CB Richard Ellis' Nick Jopling. "Accommodation would be top quality but functional, not built to Manhattan or Knightsbridge specifications. It would not incur excessive management costs, would be close to transport hubs and built in areas with an appropriate infrastructure" he says.

And why is build-to-let of interest to individual investors? Well, it will allow them to use investment of anything from £1,000 to £100,000 to be channelled into a new development via a REIT—if the REIT reforms come on stream as anticipated.

9. Rent out a garage

Bizarre but true—if you own a garage that you do not use yourself, why not rent it out? One landlord, quoted on *www.propertyhawk.co.uk*, did exactly that:

> I therefore started researching how I could rent out my garage for extra income. After committing to clearing out some of the rubbish I answered a few speculative adds in Gumtree. Within several days I'd had one reply. A lovely couple who wanted to store their Mazda Mx5 over winter. As I'm an owner of one of these classic marques I immediately took to Sarah and Ewan. We met, were happy with each other and worked out a deal of £43 a month. By searching the internet I found a free tenancy agreement for renting out my garage on a dedicated site called *www.yourparkingspace.com*. Now £43 a month or just over £500 a year doesn't seem much. However, with interest rates at 0.5% I'd have to have £100,000 in the bank to be paying a similar income. In this credit crunch environment it definitely pays to sweat your property assets.

How true ...

If you look at *www.parklet.co.uk* and go to its parking rental guide price, you can get an idea of how much a parking space costs in your area.

10. Improve your own home

This sounds like stating the obvious but everyone, especially those with only modest sums to invest, should consider whether they would get the best return by putting the funds into improving the home they live in with a view to obtaining a better price when they ultimately sell.

The single most effective way of adding value is creating an extra room (preferably a bedroom) for your house—but be careful you don't spend more than you could possibly recoup through a higher selling price.

The Federation of Master Builders says the following are the crude costs of home improvements:

1. Loft conversion to create extra bedroom—£35,000 to £45,000.
2. Sideways or rear extension—£15,000 to £25,000.
3. Basement conversion—£15,000 to £120,000.
4. New bathroom—£3,000 to £6,000.
5. New kitchen—£4,000 to £25,000.

Resource centre

The website *www.searchifa.co.uk* allows you to find an independent financial advisor in your area; the Chartered Institute of Accountants in England and Wales can be useful for information—see *www.icaewfirms.co.uk*; government websites giving tax and legal advice on renting out rooms and investing in property include *www.hmrc.gov.uk* and *www.direct.gov.uk*. You may also find the following these helpful: *www.financialombudsman.org.uk*, *www.fsa.gov.uk*, *www.olso.org*, *www.slso.org.uk* and *www.vla.nics.gov.uk*.

Appendix 1

The Assured Shorthold Tenancy Agreement

The Assured Shorthold Tenancy (AST) is almost certainly the type of tenancy you will want to have in a buy-to-let property, as long as:

- the tenant is an individual or group of individuals and not a company
- the self-contained property is let under a single agreement
- the property is the tenant's main home.

But ASTs do not apply if:

- the letting is at no rent or low rent (less than £250 per year or pro rata for just part of the year)
- the letting is at a high rent (over £25,000 per year or pro rata for just part of the year)
- the property is a holiday lettings
- the letting is where the landlord will be living at the same property
- the property comes with more than two acres of agricultural land
- tenants in situ that are currently on a protected tenancy.

Tenancies in the excluded categories will usually be "contractual" tenancies or the tenant will be granted a "licence to occupy".

The AST agreement must contain:

- the name of the landlord and a UK address for the serving of notices
- the name of the tenant
- the address of the property to be let
- the start date of the tenancy
- details of the term of the tenancy
- the amount of the rent and when it is due
- the amount of any deposit held
- if there is a guarantor, then the details may also be included in the AST or in a separate document.

If there is more than one tenant then the terms of the tenancy agreement should make it clear that they are jointly and severally liable. The tenancy agreement should be signed by all adult tenants.

An AST is for an initial fixed term. During this initial term the tenant has security of tenure, but if there are serious breaches of the tenancy agreement (such as rent arrears) the landlord can issue a Section 8 notice to demand that the tenant vacates the property even within the initial term. If the initial term is for less than six months then the tenant still has security of tenure for six months.

If the tenant wishes to end the tenancy within the term then the landlord has a duty to mitigate losses by finding a new tenant, and the tenant is responsible for rent until a replacement is found, and for any of the landlord's costs involved in finding the replacement tenant.

AST agreements are most commonly issued with a term of six months, although they can be for any length of time. Many buy-to-let mortgages have restrictions on the length of the term. For terms exceeding three years, the tenancy must be executed as a deed, which requires a solicitor.

The rent is normally payable monthly in advance, but can, if specified within the tenancy, be paid weekly or quarterly in advance or in arrears.

Notice periods can depend on the rent interval. It may be preferable to have a formal agreement, stating rent will be paid monthly and then apply an informal agreement to have rent paid according to another schedule.

An AST is essentially just an agreement permitting a tenant to occupy a property as their own home for the term of the agreement. The tenant agrees to pay the rent agreed and to return the property and the landlord's fixtures and fittings in the same condition (allowing for reasonable wear and tear) at the end of the tenancy.

The landlord will maintain the fabric of the property and service to the property. The landlord will permit the tenant to live in peace.

The detailed terms and conditions of the tenancy agreement may specify certain restrictions such as:

- preventing the tenant from decorating or otherwise altering the property
- preventing the tenant from changing gas/electricity supplier or tariffs
- clarify access rights for the landlord to inspect the property during the tenancy or carry out viewings for potential tenants towards the end of the tenancy.

Signing a tenancy agreement

The landlord should prepare two copies of the tenancy agreement. The landlord, the tenant(s) and the guarantor (if there is one) should each sign both copies. One copy is to be retained by the tenant and the other by the landlord.

At the same time as the tenancy agreement is signed, the copies of the inventory should also be signed. The inventory is a list of all fixtures and fittings along with a detailed description of the condition. This is to resolve any possible disputes over whether the contents have all been returned at the end of the tenancy and whether any decline in condition is due to damage or wear and tear. As with the tenancy agreement, all parties should sign two copies, with one retained by the landlord and the other by the tenant.

There is no legal requirement for a witness to also sign the agreement if the tenancy agreement is for less than three years.

The landlord has a right to possession of the property at the end of the term. Possession must be requested by issuing a Section 21 notice, giving the tenant two full months of notice. The landlord need give no reason for requesting possession when issuing a Section 21 notice and provided the notice is completed and issued correctly then the tenant has no grounds for appeal.

If there have been serious breaches of the tenancy agreement, such as unpaid rent, then the landlord may issue a Section 8 notice. Under Section 8 the notice period is shorter then for Section 21. The actual notice period is dependent on which of the 17 possible grounds for repossession are specified. Unlike Section 21, Section 8 can be used to evict tenants even within the term of the AST.

The tenant may also request to terminate the contract. Although the AST commits the tenant to a minimum period of the term, and if they wish to end the tenancy earlier, they may be liable for rent for the full term. If the tenant does wish to leave before the end of the term for the agreement then the landlord is advised to find a new tenant as soon as possible. The outgoing tenant will then be responsible for the rent up to the date that the new tenant moves in and for the landlord's fees in finding a new tenant.

The tenancy can also be ended by mutual consent. If no action is taken by the landlord to end the AST at the end of the term and the tenant remains at the property, then a Periodic Tenancy automatically comes into force. A Periodic Tenancy is essentially a rolling AST with notice periods of:

- two full rent periods, ending on a rent day from the landlord for the tenant to leave (using a Section 21 notice) or
- one full rent period, ending on a rent day from the tenant to end the tenancy.

Appendix 2

Tenancy Deposit Scheme

This is one area of buy-to-let that has changed substantially in recent years, and landlords who handle their own tenant- and property-management need to be aware of the varying schemes that have existed since 2007. They apply to anyone who signs an Assured Shorthold Tenancy (AST) agreement in England and Wales, and any disputes arising over return of deposits can be resolved through them.

The custodial scheme

In this, landlords pay all deposits into the Deposit Protection Scheme either online or by a telephone transaction. It is free to landlords and lettings agents as the scheme uses part of the interest earned from holding the deposit to cover costs. When it gets to the end of the tenancy, both parties have to agree on the amount of deposit to be returned in order for the scheme to release the funds. Funds are then released directly into the tenant's bank account. (More details are available from *www.depositprotection.com*.)

The insurance-based scheme

There are two options here. Landlords can use either My Deposits (*www.mydeposits.co.uk*) or the Tenancy Deposit Scheme (*www.direct.gov.uk/en/TenancyDeposit/index*). Under these options, the tenant pays the deposit to the landlord/agent who keeps the deposit during the tenancy and pays a premium to the insurer.

Landlords should choose whichever scheme they prefer and inform the tenant. If they fail to do so, and the tenant later proves to a county court judge that they have so failed, it may be difficult to serve a possession notice to ensure the tenant leaves the property and may be penalised with a fine of three times the deposit.

Appendix 3

Contract for Builder Working on Developing Property

It is perfectly sensible for anyone commissioning builders to present them with a contract—indeed, this will help future misunderstandings be dealt with professionally and quickly. (Contract courtesy of Birmingham City Council Trading Standards Department.)

CUSTOMER:

CONTRACTOR: Is it a limited company?
If not state individual's name

Name ....

Name ....

Address ....

....

Address ....

....

Tel No. (home) .... (work) .... Tel. No. ....

THE JOB:

1. Work to be done
(List all the work you have asked the contractor to carry out)
. . . .
. . . .
. . . .
. . . .
. . . .

If there is not sufficient space to list everything attach a sheet of paper.

2.   Materials to be used
(List the type, quality and measurement of materials agreed upon e.g. type of bricks, window frames, loft insulation—get samples if possible).
. . . .
. . . .
. . . .
. . . .

Note those materials to be supplied by the customer and when they are needed by the contractor

3.   Date work will commence
(Ask the contractor to give you a starting date. If it is essential that work should start on this date make a written note of this).
. . . .
. . . .
. . . .
. . . .

4.   Date work will be completed
(Ask the contractor to give a date (even if approximate) as to when the work will be finished. Again, if the work must be completed by a certain date, make a written note of this).
. . . .
. . . .
. . . .
. . . .

If the contractor fails to start or complete work by a date noted above as essential, then you may be entitled to compensation for unavoidable losses due to the delay. However this will not apply where the contractor is unavoidably delayed by circumstances either outside his control (e.g. weather) or caused by you (e.g. changing the work to be done).

5.   Will a sub-contractor be used? If so name of sub-contractor.
(Name and Address of any other contractor that may be employed by the main contractor).

. . . .
. . . .
. . . .
. . . .

THE PRICE
1.   Price Agreed
(It is essential to obtain a final fixed price or written quotation from the contractor. Remember an estimate is not binding. Check whether this price includes V.A.T.)

. . . .
. . . .
. . . .
. . . .

2.   Form of Payment
(Specify whether cash and/or credit)

. . . .
. . . .
. . . .

3.   Credit Details
(a)   Who will arrange the credit? (Specify customer or contractor)

. . . .
. . . .
. . . .
(b) Name and address of Finance Company?

. . . .
. . . .
. . . .

Never sign a credit form unless all the details have been completed.

4.   When is payment due?
(Specify when payment is to be made e.g. after completion of work or in instalments)

....
....
....
....

5.   Deposit Details
(Specify if a deposit has to be paid, if so, when and what amount? Make a note of whether or not it is refundable if the contract does not go ahead)

....
....
....
....

OTHER CONDITIONS:
1.   Cancellation Rights
(List the circumstances, if any, when you may cancel e.g. if you or the contractor are unable to arrange credit—or if you are told by the contractor there is a cancellation period) Remember you may have a right to cancel if the contractor visited your home without you asking him to.

....
....
....
....

2.   Planning Permission/Building Regulations
(Specify whether or not planning permission and/or building regulations approval is required and if necessary has it been obtained and whose responsibility is it to obtain it).

....
....
....
....

3. Special Instructions/Conditions
(List any special instructions you have given the contractor or any promises the contractor has made)

....
....
....
....

4. Access and Facilities
(Specify access times and detail those facilities which the contractor may use free of charge e.g. water, electricity, toilet).

....
....
....
....

SIGNATURES—REMEMBER A SIGNATURE IS BINDING

Customer's Signature .... Date ....

Contractor's Signature .... Date ....

.

Appendix 4

Dealing with Problem Estate Agents

If you have built your own home and now want to sell, or are disposing of a property you have redeveloped, the sale itself is the moment when you realise your profit. You do not want to jeopardise your hard work with a poor estate agent who fails to get the right buyers or the best price.

Which?, formerly the Consumers' Association, has kindly agreed that we can use the following advice note from its website, *www.which.co.uk*, on how to extract yourself from a contract with an estate agent.

If you have a 'sole selling' agreement
You must wait until the notice period elapses even if you've found a private buyer yourself.

If you have a 'sole agency' agreement
Check how long the period runs for and what notice you have to give. Avoid instructing another agent until the end of your sole agency period because if they find a successful purchaser within that time you'll still be liable to pay commission to your original agent. This means you could end up paying double commission. You can still find a private buyer yourself during that period.

How to withdraw an instruction
Call the agent. Make a note of the time and who you speak to. Tell them you wish to give notice that you're withdrawing instructions for them to sell your property,

effective immediately. Tell them you've made a note of the call and will confirm again in writing. You don't have to give them an explanation as to why you're withdrawing.

Withdrawing instructions to an estate agent to sell a property—joint agents

Your address

Dear [Reference: address of seller's property]

Further to our telephone conversation of [date], I write to give you formal notice of the withdrawal of my instructions in connection with the sale of the above property.

I understand that my sole agency period will expire on [insert date] and that I will have the right to sell my property through another agent from that date, without becoming liable to pay you any commission, subject to the terms of our contract.

Yours sincerely,

Withdrawing instructions to an estate agent to sell a property—sole agency agreement

Your address

Dear [Reference: address of seller's property]

I have received your letter of [date] requesting payment of the sum of [£......], representing [.....per cent] commission on the sale of the above property to [purchasers].

As you know, we entered into a multiagency agreement with you. We had instructed [another agent] on [date]. As you know, we entered into a multiagency agreement with you. We had instructed [another agent] on [date], prior to our placing instructions with you, and it was on [date] that [purchasers] first inspected our property.

The fact that you subsequently issued particulars of our property to the purchasers is irrelevant, since it was [other agent] who introduced them to the property which they eventually bought, prior to our placing instructions with you, and it was on [date] that [purchasers] first inspected our property. The fact that you subsequently issued particulars of our property to the purchasers is irrelevant, since it was [other agent] who introduced them to the property which they eventually bought.

Yours sincerely,

Appendix 5

Home Information Packs

Whether your home is self-built, newly redeveloped or a buy-to-let, you will realise much (perhaps all) of its profit only when you sell. Most of us are familiar with selling through estate agents but few long-term investors may have yet dealt with the Home Information Pack (HIP) which is now compulsory on all sales.

You don't need a HIP for:

- properties where there is no marketing, for example if you are selling to a member of your family
- non-residential properties
- properties limited by law to use as holiday accommodation or occupation for less than 11 months per year
- mixed sales, for example a shop with flat
- sales of portfolios of properties, for example selling two properties together
- properties not being sold with completely vacant possession, for example with a sitting tenant
- unsafe properties and properties due to be demolished
- properties sold through the "Right to Buy", "Right to Acquire" and "Social HomeBuy" home ownership schemes.

You DO need to provide a HIP for newly built properties and all residential properties that do not fall into the categories set out above. Before you even begin marketing a property, your HIP must be compiled and consist of a HIP index. The contents are listed below, courtesy of the Association of Home Information Pack Providers:

PART 1—General—Required Documents

Please look at each document listed in column 1 and then complete the relevant entry in either column 2 or column 3

Column 1 **Home Information Pack document**	Column 2 **Included** **Date on document and any further information**	Column 3 **If it is a required document for your property:** • Confirmation that proof of the request for the document is included (for documents required within 28 days of marketing); • reason why not included; • steps being taken to obtain it; • date when it is expected to be obtained; • any reason for further delay and further date by which the document is expected.
1. Index		
2. Property Information Questionnaire		
3a. Energy Performance Certificate and Recommendation Report—or:		
3b. Predicted Energy Assessment		
4. Sustainability Information (New Homes only)		
5. Sale statement		
Title information		
6. Official copy of the individual register (for registered properties only)		
7. Official copy of the title plan (for registered properties only)		

8. Certificate of official search of the index map (for unregistered properties only)	
9. Documents provided by seller to prove title (for unregistered properties only)	
10. Leases, tenancies or licences for dwellings in a sub-divided building that are being marketed as a single property and where part of the property is being sold with vacant possession	
Search reports	
11. Local land charges	
12. Local enquiries	
13. Drainage and water enquiries	
PART 2—Commonhold Properties—Required Documents	
1. Land Registry individual register and title plan for common parts	
2. Land Registry copy of commonhold community statement	
3. Management rules and regulations outside the commonhold community statement	
4. Requests for payment towards commonhold assessment for the past 12 months	
5. Requests for payment towards reserve fund for the past 12 months	
6. Requests for payment towards insurance for common parts for the past 12 months (if separate to commonhold assessment or reserve fund)	
7. Name and address of managing agents and/or other manager (current and any proposed)	
8. Amendments proposed to the commonhold community statement, and other rules	
9. Summary of works affecting the commonhold (current and any proposed)	
10. Where the commonhold interest has not been registered at the Land Registry: the proposed commonhold community statement and an estimate of costs expected of the unit-holder in the first 12 months	

PART 3—Leasehold Properties—Required Documents
Please look at each document listed in column 1 and then complete the relevant entry in either column 2 or column 3

Column 1	Column 2	Column 3
Home Information Pack document	**Included** **Date on document and any further information**	**If it is a required document for your property:** • Confirmation that proof of the request for the document is included (for documents required within 28 days of marketing); • reason why not included; • steps being taken to obtain it; • date when it is expected to be obtained; • any reason for further delay and further date by which the document is expected.
1. The lease, being either: • an "official" copy; • the original lease or a true copy of it; or • an edited information document.		

PART 4—Authorised Documents	
Home Information Pack document	**Included** **Date on document and any** **further information**
Please list any authorised documents that have been included relevant to this property below:	
1.	
2.	
3.	
4.	
5.	
6.	
7.	
8.	
9.	
10.	
11.	
12.	
13.	
14.	
15.	
16.	
17.	
18.	
19.	
20.	

In addition, you must complete a Property Information Questionnaire (PIQ) and, depending on whether your property is second hand or new, either an Energy Performance Certificate (EPC) or a Predicted Energy Assessment (PEA).

All of these—the HIP Index, the PIQ, EPC and, if applicable, PEA —must be in place before you even market the home.

If the following documents are unavailable when marketing begins, they should be added to the HIP as soon as they are available, and must be included within 28 days of the date the property was first placed on the market:

- Standard searches (local authority and drainage and water).
- A copy of the lease for leasehold properties.
- Commonhold documents, where appropriate.

If you do not wish to do this yourself, a HIP can be provided by an estate agent, a solicitor or a specialist HIP provider—scouring the internet will find ones close to your property for sale.

Appendix 6

Types, Classes and Uses of Commercial Property

The Town and Country Planning (Use Classes) Order 1987 (SI 1987/764) is reproduced below:

SCHEDULE
PART A

Class A1. Shops
Use for all or any of the following purposes—

(a) for the retail sale of goods other than hot food,
(b) as a post office,
(c) for the sale of tickets or as a travel agency,
(d) for the sale of sandwiches or other cold food for consumption off the premises,
(e) for hairdressing,
(f) for the direction of funerals,
(g) for the display of goods for sale,
(h) for the hiring out of domestic or personal goods or articles,
(i) for the reception of goods to be washed, cleaned or repaired,

where the sale, display or service is to visiting members of the public.

Class A2. Financial and professional services
Use for the provision of —

(a) financial services, or
(b) professional services (other than health or medical services), or
(c) any other services (including use as a betting office) which it is appropriate to provide in a shopping area,

where the services are provided principally to visiting members of the public.

Class A3. Food and drink
Use for the sale of food or drink for consumption on the premises or of hot food for consumption off the premises.

PART B
Class B1. Business
Use for all or any of the following purposes—

(a) as an office other than a use within class A2 (financial and professional services),
(b) for research and development of products or processes, or
(c) for any industrial process,

being a use which can be carried out in any residential area without detriment to the amenity of that area by reason of noise, vibration, smell, fumes, smoke, soot, ash, dust or grit.

Class B2. General industrial
Use for the carrying on of an industrial process other than one falling within class B1 above or within classes B3 to B7 below.

Class B3. Special Industrial Group A
Use for any work registrable under the Alkali, etc. Works Regulation Act 1906[5](a) and which is not included in any of classes B4 to B7 below.

Class B4. Special Industrial Group B
Use for any of the following processes, except where the process is ancillary to the getting, dressing or treatment of minerals and is carried on in or adjacent to a quarry or mine:—

(a) smelting, calcining, sintering or reducing ores, minerals, concentrates or mattes;
(b) converting, refining, re-heating, annealing, hardening, melting, carburising, forging or casting metals or alloys other than pressure die-casting;
(c) recovering metal from scrap or drosses or ashes;
(d) galvanizing;
(e) pickling or treating metal in acid;
(f) chromium plating.

Class B5. Special Industrial Group C

Use for any of the following processes, except where the process is ancillary to the getting, dressing or treatment of minerals and is carried on in or adjacent to a quarry or mine:—

(a) burning bricks or pipes;
(b) burning lime or dolomite;
(c) producing zinc oxide, cement or alumina;
(d) foaming, crushing, screening or heating minerals or slag;
(e) processing pulverized fuel ash by heat;
(f) producing carbonate of lime or hydrated lime;
(g) producing inorganic pigments by calcining, roasting or grinding.

Class B6. Special Industrial Group D

Use for any of the following processes:—

(a) distilling, refining or blending oils (other than petroleum or petroleum products);
(b) producing or using cellulose or using other pressure sprayed metal finishes (other than in vehicle repair workshops in connection with minor repairs, or the application of plastic powder by the use of fluidised bed and electrostatic spray techniques);
(c) boiling linseed oil or running gum;
(d) processes involving the use of hot pitch or bitumen (except the use of bitumen in the manufacture of roofing felt at temperatures not exceeding 220°C and also the manufacture of coated roadstone);
(e) stoving enamelled ware;
(f) producing aliphatic esters of the lower fatty acids, butyric acid, caramel, hexamine, iodoform, napthols, resin products (excluding plastic moulding or extrusion operations and producing plastic sheets, rods, tubes, filaments, fibres or optical components

produced by casting, calendering, moulding, shaping or extrusion), salicylic acid or sulphonated organic compounds;

(g) producing rubber from scrap;
(h) chemical processes in which chlorphenols or chlorcresols are used as intermediates;
(i) manufacturing acetylene from calcium carbide;
(j) manufacturing, recovering or using pyridine or picolines, any methyl or ethyl amine or acrylates.

Class B7. Special Industrial Group E
Use for carrying on any of the following industries, businesses or trades:—

Boiling blood, chitterlings, nettlings or soap.

Boiling, burning, grinding or steaming bones.

Boiling or cleaning tripe.

Breeding maggots from putrescible animal matter.

Cleaning, adapting or treating animal hair.

Curing fish.

Dealing in rags and bones (including receiving, storing, sorting or manipulating rags in, or likely to become in, an offensive condition, or any bones, rabbit skins, fat or putrescible animal products of a similar nature).

Dressing or scraping fish skins.

Drying skins.

Making manure from bones, fish, offal, blood, spent hops, beans or other putrescible animal or vegetable matter.

Making or scraping guts.

Manufacturing animal charcoal, blood albumen, candles, catgut, glue, fish oil, size or feeding stuff for animals or poultry from meat, fish, blood, bone, feathers, fat or animal offal either in an offensive condition or subjected to any process causing noxious or injurious effluvia.

Melting, refining or extracting fat or tallow.

Preparing skins for working.

Class B8. Storage or distribution
Use for storage or as a distribution centre.

PART C

Class C1. Hotels and hostels
Use as a hotel, boarding or guest house or as a hostel where, in each case, no significant element of care is provided.

Class C2. Residential institutions
Use for the provision of residential accommodation and care to people in need of care (other than a use within class C3 (dwelling houses)).
Use as a hospital or nursing home.
Use as a residential school, college or training centre.

Class C3. Dwellinghouses
Use as a dwellinghouse (whether or not as a sole or main residence) —

(a) by a single person or by people living together as a family, or
(b) by not more than 6 residents living together as a single household (including a household where care is provided for residents).

PART D

Class D1. Non-residential institutions
Any use not including a residential use —

(a) for the provision of any medical or health services except the use of premises attached to the residence of the consultant or practioner,
(b) as a crèche, day nursery or day centre,
(c) for the provision of education,
(d) for the display of works of art (otherwise than for sale or hire),
(e) as a museum,
(f) as a public library or public reading room,
(g) as a public hall or exhibition hall,
(h) for, or in connection with, public worship or religious instruction.

Class D2. Assembly and leisure
Use as —

(a) a cinema,
(b) a concert hall,
(c) a bingo hall or casino,

(d) a dance hall,

(e) a swimming bath, skating rink, gymnasium or area for other indoor or outdoor sports or recreations, not involving motorised vehicles or firearms.

©Crown Copyright 1987

Index

Printed and bound by CPI Group (UK) Ltd, Croydon, CR0 4YY

21/10/2024

01777049-0002